WILD HORSES, WILD WOLVES

WILD HORSES, WILD WOLVES

Legends at Risk at the Foot of the Canadian Rockies

MAUREEN ENNS

RMB

Gathering Scent, mixed media painting, 48 × 36 in., Enns, 2008.

CONTENTS

EL CID ON CAMP RIDGE, 2012.

FOREWORD

It is nice to know there are still mysteries close to home that cannot be answered with an Internet search, that in fact simply do not have a "for sure" answer. Within one hundred kilometres of a city with over one million people, there are magnificent stallions running with small herds of mares and foals that very few people are aware of. Not even the Natives at nearby Rabbit Lake, nor the adjacent rancher on the Lazy JL Ranch who runs cattle in the same valley during the summer, nor the recreational users on their quads and motorcycles who tear through the terrain, nor the men who service the natural-gas wellheads, and not even the conservation officers charged with managing the land, have any idea how many bands of wild horses are resident in this small foothills habitat.

Where did these magnificent proud stallions with manes flowing down to their knees, and tails brushing the ground, come from? Have they been here forever? Did they emerge from Native legends? Are they descended from Spanish stock come north with the settlers and Natives across North America? Or are their origins more mundane? Are they the castoffs from ranchers and equestrians who could no longer keep them but could not kill them? Was it the transition to the internal combustion engine or the economic collapse followed by the Dirty Thirties that prompted their release? Have they been running free for a century, or is this a more recent phenomenon?

Does it matter where they came from in order to justify their presence on the landscape?

What we do know – thanks to over six years of Maureen Enns's devotion and personal research – is that the horses are there and they do not need us to help them survive. They are wild and free and fitting into a habitat niche that seems to suit them very well.

They co-exist with cattle, elk, moose, deer, bears, wolves, coyotes and cougars, and nowhere is the range depleted or overgrazed. Can these relationships between creatures and land best be described as a balance, or as a pendulum that swings back and forth over a time frame that we do not perceive?

You may never see these horses, for somewhere in their past they have learned that survival depends upon avoidance of the ultimate predator – man. Over the years some have been baited and tricked into capture corrals, where they have been subdued and broken to stock horses or shipped south to the meatpacking plant for overseas consumption. Fortunately, the market is currently weak for both products, and trapping no longer has a significant impact on their numbers.

Their adaptive behaviour is strikingly similar to the deer who share the forest landscape with them: "Freeze: don't move and you won't be seen." That is, until the flies are so bad that your tail has to twitch or some part needs to be scratched. Movement is the giveaway. Maureen has never been known to willingly sit still for any length of time, but to know the "wildies," one must meet them on their terms. "Patience, Grasshopper. Patience."

The horses do not exist in a national park, not even a provincial park. Their range is the Ghost Forest Land Use Zone (FLUZ), a multidimensional area of integrated land uses. Almost anything is permissible there: oil and gas exploitation, timber harvest, cattle grazing, hunting (including wild-horse capture), fishing, camping, all-terrain vehicles, motorcycles and snow machines. Have we missed anything? And yet a complex web of wildlife thrives under the radar.

◈ ◈ ◈

I first met Maureen Enns in 1991, when I was the senior wildlife warden in Banff National Park. She arrived in my office with a proposal to study, and ultimately paint, grizzly bears. At first she was just another of the long list of filmmakers, photographers, journalists and artists who come to the national parks in need of access and assistance to interpret nature and hopefully produce a work of art. But this lady was different from the others. She brought her own horses and could throw a diamond hitch. She was self-sufficient and mostly just needed permission to carry on. She also had a background in conservation-related art and was a recognized instructor at the Alberta College of Art.

Maureen knew that the grizzly was misunderstood, and she spent three summers in pursuit of the elusive animals in Banff National Park.

When asked to write the introduction to her book *Grizzly Kingdom* back in 1994, I wrote the following:

> … She is one of the few persons who has taken on a topic as highly polarized as the survival of the grizzly and managed to treat it impartially from all angles. Ranchers, hunters, environmentalists, researchers, trainers, tourists, wildlife managers and many others all had input and were represented in her work.
>
> The survival of the grizzly lies in habitat protection and understanding, not isolation. Maureen's works have brought the grizzly to countless people who might not otherwise have encountered the great bear and given it any thought.

Twenty years later these same words ring true; just replace "grizzly bear" with "wild horse."

In the intervening years, Maureen spent nine summers researching and living with the grizzly bears with Charlie Russell on the Kamchatka Peninsula in eastern Russia. Raising three orphaned grizzly cubs, and babysitting for the local bears, allowed her to gain insights into behavioural communication not accomplished by others.

Maureen and Charlie were accepted into the midst of the greatest concentration of wild grizzly bears anywhere in the world, proving that understanding is the key to survival. Their work can be seen in the award-winning documentary film *Walking with Giants* and read in their book *Grizzly Heart*.

Maureen has always had a passion for creatures that are wild and free. One look at her mane and it is quite obvious there is a common bond between her and horses. She grew up on horseback, subduing them to her need to run free and jump. In her younger years, she took many equestrian titles in western Canada. It is only fitting that now, somewhat more mature, she perceives the need to preserve wild horses' freedom and right to run and jump.

Her art has captured many of the quintessential movements and expressions that only wild animals are capable of. I urge you to follow Maureen through the following pages as she unfolds the mysteries of the Ghost Forest and sheds light on a complex interplay of critters that are surviving wild and free despite our encroachments into their habitat.

— Rick Kunelius
Wildlife Conservation Consultant, Banff, Alberta

KIT'S BAND, 2010.

ACKNOWLEDGEMENTS

Thinking of Rick Kunelius, who wrote the foreword to this book, I am grateful for his continued faith in my work with misunderstood animals and his help and encouragement.

Eleanor O'Hanlon, an Irish writer living in France, while conducting research for her latest book, *Eyes of the Wild: Journeys of Transformation with the Animal Powers*, found me on the Internet. She was fascinated by my relationship with grizzly bears in Kamchatka and the wild horses of the Ghost Forest in the Alberta Rockies foothills northwest of Calgary. I would not have written this book without her appreciation of my observations of the wild horses of my study area. Yes, it took someone from far away who is held in high regard in conservation circles in Europe to tell me I was onto something.

I thank Penny MacMillan and our friends Arndt Salinsky and Marina Krainer of the Ghost Watershed Alliance Society, who told me about the wild horses in the southeast corner of the Ghost Forest in 2006. Penny's help and friendship throughout these past six years of hard research has been invaluable.

I wish to thank Roger Vernon for his assistance with camera installations and support throughout. Brenda Gladstone of the Galileo Educational Network and the folks at Pyramid Productions film company in Calgary recognized the value of my work, which gave me the necessary boost in confidence with so much adversity to wild horses in Alberta. Rod and Lois Green of Masters Gallery Ltd. in Calgary seemed to always have an art sale of my work when funds ran low. Their interest in conserving the area and the rewilded horses was critical at several stages, as was their friendship and Lois's development of sculpture featuring the wild horses of the Ghost Forest. I thank Cori Brewster for her interest in the wild horses and commitment to writing the lyrics and the music about them.

Many ranchers and people in the horse industry helped with encouragement and with permission to interview them for the historical content of this book. I am grateful for information from Donna Butters, Erin Butters, Alyssa Butters and Darcy Scott. Without the help of Erik Butters and his shared stories and permission to reproduce family photographs, this book would have lacked historical character. The Butters clan are a wonderful family and ranchers of the highest calibre. Thanks to Hamish Kerfoot of Providence Ranch for his story of the grand escape of fine Kerfoot horses in the late 1920s, as well as the use of his important family photos. Others who generously

Rick caught in perfect light.

Penny on her gelding, overlooking Deer Lake marsh and the Devil's Head mountain.

offered their time, sharing important knowledge about wild horses, include Pam Asheton, Chuck and Terry McKinney, and Dr. Claudia Notzke. I thank "the trapper," who shall remain anonymous throughout, for his time and open disclosures about his work..

Thanks also go to Butters Ranching Ltd. for giving me unlimited permission to cross their ranch and access their grazing allotment, where the wild horses live. I thank Anita and Cody Wildman for giving me permission to cross by their house on a regular basis on the Rabbit Lake Indian Reserve en route to Owl Creek Road. I also thank Direct Energy for allowing me to park near their well site, which offered some protection to my rig when

I was out all day. Thanks go to former Alberta Minister of Environment and Sustainable Resource Development Ted Morton for meeting with me on more than one occasion to discuss the possibility of a wild-horse conservation area in the Ghost Forest. I thank Ken McKay of Alberta Fish and Wildlife for his interest in my project and his ongoing encouragement.

I offer a heartfelt thank you to Anita Crowshoe, who generously shared two traditional oral stories of the Piikani Nation and offered wonderful support when I was tired of writing my story. Thank you, Dr. Brian Reeves, for sharing your historical knowledge as an archaeologist regarding the first wild-horse remains and other aspects of how horses came to the Ghost Forest. I am grateful for the help of Peter Snow and his uncle, Hank Snow, for their support and critical assistance while I was gathering stories from the Stoney Nakoda Nation elders for the *Wild and Free* educational initiative.

I thank the former superintendent of Banff National Park, Kevin Van Tighem, who introduced me to Don Gorman, publisher of Rocky Mountain Books. Don, thank you for your immediate interest in my research and for agreeing to publish and promote my manuscript and photography. Thanks go to Nancy Townshend for her enthusiastic suggestion that I contact Meaghan Craven for editorial assistance; Meaghan has turned out to be a gem from Nelson, BC. Finally, thanks to the staff at RMB | Rocky Mountain Books: Chyla Cardinal, art director; Neil Wedin, marketing coordinator; and Joe Wilderson, senior editor.

❋ *HIDING BEHIND ONE TREE,* CHARCOAL DRAWING, ENNS, 2007. ❋

CHAPTER ONE

From Grizzly Bears to Wild Horses

In the late summer of 2006 I forced myself to go for a ride at the first light of day, around 6 a.m. In spite of consuming extra coffee on the drive up Owl Creek Road with my mare, Hope, in the trailer, I was still half-asleep by the time I reached the edge of Deer Lake marsh an hour later. For two weeks I had ridden into the area with no sightings of wild horses, but I was sure I would see one this morning with all the fresh signs on the trail.

Suddenly a herd of horses – three black and three bay – galloped toward us. They were all wild stallions. I had read that stallion colts in wild herds leave the family group at about four years to wander alone for a while or join a group of young bachelors. However, my first thought that morning was not about wild horses. Instead, I thought the six stallions were domestic. After all, they did not run away. My second thought was, "Oh, shit. Now I'm in trouble." I worried they would run up to us, as domestic herds do, try to sniff us, snort, bite, squeal and kick, compete for the pretty chestnut mare. I knew I was in a dangerous spot astride Hope, who I could feel was becoming excited. She twitched her tail – which is rather like winking – indicating her readiness to breed.

I wondered if one would try to mount Hope with me on her back.

To my astonishment the herd came to a sliding stop, the six of them attempting to hide behind some trees. They all chose one spindly tree at the same time, colliding in their confusion. One edged around to look at us, then the others followed suit, shoulder-to-shoulder as if playing peek-a-boo. I could imagine one telling the other, "You go forward!" And the reply, "No, you!" But not one young stallion moved out of alignment. Their antics reminded me of my new blue heeler pup, Spook, who although at three months was schooled to sit and stay would always silently wiggle forward to investigate. I laughed out loud at the horses' antics, but even in the throes of the moment, I managed to take a blurred photograph of the spectacle.

Suddenly, as one, they galloped off, only to stop about ten metres away to watch us again, this time each one behind his own tree. Hope and I were riveted to the spot. It was a magical moment in time when six young studs, Hope and I curiously sized one another up. Eight pairs of different eyes connected at will.

Stunned by the display of respectful curiosity, I almost

Bachelor herd checks us. Note the injury to the black, which likely happened due to an unchecked run through the forest. He received no vet care, yet he survived.

Devils Head Meadows, painting on canvas, 48 × 60 in., Enns, 2006.

Bachelors Gallop Off as One, charcoal drawing, 21 × 30 in., Enns, 2006.

forgot to take more blurry photos before they ran away, hurtling forward, then circling to grab another furtive look as they flashed by in a blur of black and brown, manes flying, tails aloft. They made one more pass, and then, hearing the beat of their leader, they wheeled and were gone as one muscled mass of power and grace.

I realized that after my initial surprise, I no longer felt fear of the herd. I have learned to trust my horse and know she can communicate with other animals in ways I cannot. I trusted her assessment of the stallions that day as I had trusted the intuition of my gelding, Spud, when we encountered a mother grizzly over a decade before.

If one is lucky, great epiphanies happen once or twice in a human's life. For me, this was one of them. After this chance encounter with the bachelor herd, a drawing and a painting poured from my body without thought or meaning – only excitement. When rational thought surfaced, I realized I had crossed the boundary; I knew the horses I met in the Ghost Forest were not domestic, nor were they just domestic horses somehow making it in the wild – they were different.

My foray into the world of wild horses, which you will encounter in this book, is part of a longer process for me that has involved the close study of wild and domestic animal behaviour. I have journeyed long to arrive at the Ghost horses of the Rockies, and the start of my experience began in a remarkable way, with my study of grizzlies in the Canadian Rockies and then on the Kamchatka Peninsula.

◈ ◈ ◈

In the 1970s I purchased some mountain ponies and a beat-up truck and horse trailer. I asked many a cowboy about their encounters with grizzlies in the Canadian Rockies, as I was eager to ride in the backcountry and a bit nervous about the bears. Without exception, their responses fed my fear. "First smell of a griz and your horse will dump you!" or "They are sneaky. I've had one follow me for miles." Or "Remember the two women who were pulled out of their sleeping bags, dragged off and killed?" Such words of wisdom and advice were common. And so, when more than twenty years later, as an artist and photographer, I entered my study of grizzly bears, I still held on to those thoughts.

Before I began my grizzly studies, I was in Kenya, working on a series of drawings, paintings and ultimately a documentary (*Game's End*) about the demise of the African elephant during the height of the poaching wars. Keen on doing something closer to home, I dreamt up some artistic research for myself, focusing on the king of the Canadian Rockies – the grizzly bear. For three summer seasons, I photographed, drew and painted the elusive animal. The result was *Grizzly Kingdom*, a book, art exhibition and later a television documentary. It told the story of how I came to realize the grizzly was a misunderstood animal. Two significant events took place in the course of that project that changed the way I view common preconceptions about animals. Both incidents happened when I was with my horse, Spud.

In 1991, after a long day of filming, I experienced something close to mystical when riding Spud and leading Chief, my packhorse, down Cascade Fire Road in Banff Park. Both horses' heads swivelled simultaneously to the grassy area on our left. I knew immediately that some kind of game was nearby. A silvertip mother grizzly and her yearling cub were digging up roots, less than ten metres away. ("Silvertip" is a common word used in the Canadian Rockies to describe the white guard hairs that flow over the dark undercoat of a grizzly bear.)

I reined Spud in suddenly, causing Chief to bump rudely into his rump. Both horses were all eyeballs on the bears. The mother bear drifted closer, chomping grass as she progressed. Spud and Chief relaxed and took the opportunity to doze off but not before raising their tails with the resulting plop, which startled the cub. It stood and woofed, but seeing its mother unalarmed, it resumed a position on all fours and chewed grass, slowly advancing closer to us. When the cub was about three metres away, I decided to edge the horses sideways a bit as I didn't really want the cub underfoot. The shuffling action on the gravel stopped its advance, and the cub moved back in the direction of its mother, continuing to chomp grass. The five of us stayed relatively close to one another for the next fifteen minutes, at which point I finally remembered to take out my camera. I had been so excited that I had forgotten to take the pictures I would need for my studio work.

Eventually the mother bear decided to move on. With her cub obediently close at heel, she crossed in front of us, pausing with one paw held aloft before disappearing into the thicket. I was in awe. The encounter could only have happened, I think, because I was there at that moment with Spud and Chief, who had understood the situation intuitively and immediately responded to mother and cub with calmness and trust.

A month later I completed a painting titled *Queen of the Rockies*, full of the colour and supreme elegance of the female summit predator that trusted my horses and me. In my painting I wanted to convey how I held my breath as my former beliefs about horses and bears tumbled. In my painting the mother silvertip grizzly pauses with paw raised, and colours skyrocket, forming rainbow arches with flung paint and line.

On another trip into Banff National Park, I left Chief at the Divide Pass warden cabin corral and rode up to the pass between Red Deer River and Clearwater River in the northeasterly section of the park. Spud, breathing heavily, stopped suddenly and pivoted his head to the left. A mother grizzly was on an elk she had recently killed, two cubs of the year at her side. Before I could decide what to do to avoid conflict – knowing how protective bears can be over their kills – the two youngsters scampered around Spud and me and to my horror stopped about ten metres to our right, putting us between the mother and her cubs, as she hovered over fresh meat. Spud relaxed, and, as on other occasions, I trusted his animal ability to emotionally read the situation accurately. Instead of charging us as I had heard always happens in such situations, the mother grizzly calmly walked around us, collected her cubs, slid down an embankment and disappeared from sight.

Those two events with Spud mark when all the old negative mythologies about grizzlies began to unravel for me. Spud, my amazing horse who taught me so much about animal language and trusting grizzlies, died in the late 1990s – the best horse in the Rockies.

Today, Hope has taken his place as my wild-animal guru. Her area of expertise is wild horses, having roamed amongst them herself as a young range horse in the Chilcotin country of British Columbia. Hope can read the intent of a wild horse in ways I cannot imagine. I have been learning from my horses for years now. These domesticated creatures have the latent ability to relate directly and emotionally to other creatures. Of course, in the wild, this skill is paramount to survival for all animals, from grizzlies to wild horses, as I discovered when I continued my studies of grizzlies, this time in one of the most inhospitable locations in the world: the southern tip of the Kamchatka Peninsula, in Russia's Far East.

◈ ◈ ◈

For nine summers (1994–2003) I lived on Kamchatka. At the beginning of this adventure, my research partner, Charlie Russell, and I had hijacked three orphaned grizzly bear cubs from the Petropavlovsk Zoo, planning to reintroduce them into the wild around Kambalnoye Lake. The cubs – Chico, Biscuit and Rosie – were my teachers. It took me nine years living with them and the other bears of Kambalnoye Lake to interpret their body language, such as the non-aggressive signals that had caused Spud to relax naturally with the mother grizzly and cub years previously.

During the Kamchatka years, I saw that when bears are confronting one another and want to avoid a fight, they rotate from a frontal head-on position to one with their shoulders to one another, often eating grass to display relaxation. I analyzed film of that behaviour, concluding the bears' display of a sidelong view of their bodies was the key to diverting aggression. Knowing about this body language has helped me understand the relationship between wolves and wild horses in the Ghost Forest a decade later.

It took me two years to understand the body language of Rosie, always the third orphan cub in line when walking along a trail across the tundra. I followed behind the three cubs, last in line, dressed in high hip waders of Russian design (definitely a fashion statement, but one that was lost on Rosie, who was in front of me). I initially believed Rosie was last because she was lazy and of an artistic temperament, wanting more freedom to explore things along the way. Later, I realized there was another reason for the cubs' positions on the trail. Biscuit, who lived the longest, was like the mother to Rosie and Chico, and she was always in the middle of the lineup. Chico, the leader, was always in front. Rosie was last. I eventually discovered Rosie's position was rear guard, which was critical when the wind blew toward them. Something coming up from behind would be neither smelled nor heard. Me clomping

Paddling with the cubs, Kamchatka.

Maureen and Biscuit, Kamchatka, 2009. Photo by Charlie Russell.

along behind in ill-fitting rubber boots was noisy, and my scent had something of a camouflaging effect on an otherwise fresh wind blowing into our backs. As soon as I realized what Rosie's role was, I slipped into place between Rosie and Biscuit and felt a collective sigh of relief from the bears. My place was within the bear procession, not at the end! Whenever their order shifted, they knew something was wrong and would immediately stand up on their hind feet to see what it was.

While working with the bears, I researched and produced a series titled *The Bears Who Looked for Beauty*. During the summer of 1997, I sought vantage points where, inevitably, I located a bear bed nearby. I prefer to call these nesting sites, as they reminded me, in their perfect symmetry, of bird nests. By watching the cubs' behaviour at these sites, I accepted the possibility that bears have an aesthetic sensibility. They slept; they sat up to look around at whatever was moving; they relaxed and simply gazed into space, seemingly admiring the view.

Upon the conclusion of my nine-year study of the Kamchatka grizzlies, I tagged an expression to summarize how I view nature: through the eyes of the bear. This way of seeing is not so much about how I view the forest or track an animal – most hunters can do that. It is about how I seek to interpret subtleties in animal communication. Bears do that. Leaving Kamchatka in 2003, I realized I had ratcheted up a notch the possibility of humans and grizzly bears co-existing peacefully. What the future had in store for me was impossible to say – I would not have guessed wild horses.

❖ ❖ ❖

I cashed in on my pension plan from the Alberta College of Art in the early 1980s to purchase the land where I now have my studio on the Ghost River, north of Cochrane, Alberta, less than an hour's drive from Calgary. As I write this, I hear the odd rustle of bats in the eaves outside my office on the studio's second floor. On one hundred acres of land I keep two horses, my dogs Cub and Spook, three cats and eight chickens. I share my life with Rick Kunelius, who, besides being good looking and tons of fun, is full of insightful knowledge of wild animals gleaned from years as a wildlife warden in Banff National Park. Cougars, black and grizzly bears, moose, deer, lynx, coyotes and wolves pass through my land – some with greater frequency than others.

A few years ago, a cougar attacked my dog near the back porch. I leapt off the porch and kicked it, yelling obscenities. Anyone who has seen me really mad was not surprised to hear that the cat dropped my dog and fled. Instead of seeking retribution from the Department of Fish and Wildlife, I let it go, as I did when a cougar killed my prize four-month-old filly. Interfering with nature is not what I am about. I accept that "wild" nature is about survival, and it is not always kind.

During the fifteen years I studied grizzly bears, I wrestled with the definition of "wild" as it relates to animals. The word emerged in my thinking as synonymous with freedom, surviving independent of man. I have long appreciated in humans and domestic animals a strong streak of independence. My way of interpreting "wild" and "domestic" turned out to be the key to my wild-horse research.

I grew up seeing the old broncobuster's approach to killing a colt's spirit in order to gain its submission. The great horse trainers of our time – Tom and Bill Dorrance, founders of the natural horsemanship movement; Monty Roberts, Buck Brannaman and Ray Hunt among others – are aware of equine boundaries and how horses use body language to communicate with one another. Their methodology involves forming equine partnerships, not "breaking" horses. Following this philosophy of horse-human relationships, I have studied the art of dressage to refine my ability to communicate with horses. Dressage is an elegant but controlled way of riding; you use subtleties of balance and movement to talk to your horse.

Learning dressage helped me form the foundation for how I ride today and has allowed me a window through which to observe the wild herds of the Ghost Forest, their behaviour and means of communication. At the same time, my understandings of grizzly bears have offered me a doorway through which to enter the wild horses' world and understand what I see.

❖ ❖ ❖

Since that first encounter with the bachelor herd in 2006, I have discovered and studied seven herds of wild horses that have lived many generations – since the early 1900s – in their wild marshland near the Ghost River. I have learned much. Even so, I did not want to write this book during the summer of 2012. The days were long and sunny and the temperatures held in the mid-twenties – perfect for hiking or riding in the mountains. However, it was brought to my attention by my friends in the Ghost Watershed Alliance Society, Arndt Salinsky and Marina Krainer, that my idyllic six years of studying the wild herds could abruptly end. A clash looms on the horizon, and wild horses on public land will be affected. Increasing recreational use of ATVs (all terrain vehicles) – along with their impact on cattle grazers – as well as the imminent harvesting of public forestlands for timber, threaten the horses' survival.

Combined with the warning that the Ghost horses may not have much more time to roam their territory on Crown land came a realization that also helped propel me toward writing this book sooner rather than later. In the early morning hours in April 2012, six years after I'd begun studying the herds, it came to me that I had uncovered the definition of the true wild horse, the mystery of the naturized horse. This may seem like a simple statement, but it is loaded with meaning. The true wild horse is defined by its independence of humans, its wild behaviours and its ability to form relationships with other wild animals. And so I began writing my story, which takes place within the small, geographically remote wilderness area within a two-hour drive from Calgary – an anomaly in public land-use zones in Alberta.

The area the horses inhabit includes the grazing lease of a local rancher, natural gas extraction sites, public forest and most recently a trail system for ATVs. It is the home of small herds of wild horses, wolf packs, foxes, grizzly and black bears, mule and white-tailed deer, elk, moose, ducks, loons, geese, sandhill cranes and a complement of shore birds.

Understanding the wild horse began with realizing some of their behaviour was similar to that of deer. My research ended with the understanding that the wild horses are part of a complex matrix of wild animals, and that in order to survive, the horses needed to form relationships with wolves, deer, moose and sandhill cranes, among others. With this research, I humbly hope to put forward a revised definition of the Ghost Forest's wild horses, one that does not include the catch-all word "feral," synonymous with "pest" in the circles of Sustainable Resource Development in Alberta. These are horses that have gone "back" to nature; I call the process "rewilding" or "naturizing."

It takes several generations for a horse to rewild, and some are more capable of that process than others – it would seem that their genetics need to be so inclined. It took time, living with the wild horses and having some experience with domestic horses, to realize the differences between a herd that has lived independent of humans for generations and a domestic herd. I have taken that time and have seen a depth of emotion and intra-herd bonding, as well as discipline and caretaking, that I would not originally have thought possible. And so the following story of the Ghost horses details the behaviours and relationships I have observed these past six years. The depth of the bonding and levels of caregiving within wolf packs has been documented by many. I have seen examples of caregiving within wild-horse family groups that equal the depth of wolf family interactions. I have also documented some unexplainable interactions between wild horses and a wolf pack that elucidate what it means to be a wild animal, dependent on other wild animals.

◈ ◈ ◈

However, research from the field is not the only piece vital to the horses' story and hopefully their conservation: stories and history are important, too.

When I moved to Alberta in the 1970s I heard Andy Russell, a well-known author, broadcast stories of importance to the history of Alberta. His radio program was called *Our Alberta Heritage.* I hold Andy in high regard and knew him better than most, having visited the Russell ranch, Hawk's Nest, while working with his son, Charlie. I miss Andy's laugh when he launched into one more story on the porch of his home near Waterton Park, repeating at least once a former radio episode. With Andy in mind, therefore, I know the wild horses' story is important. Through the stories of my ranching neighbours, the history of the horses unfolds in coming chapters, showing how they became part of our Alberta heritage.

I also decided to weave in the oral histories of Stoney Nakoda Nation elders and of my friend Anita Crowshoe of the Piikani Nation in southern Alberta, as well as a story from the scientific viewpoint of a respected archaeologist, Dr. Brian Reeves.

Finally, although these frameworks of "naturized" behaviour and history guided me as I wrote through the warm summer days of 2012, what really drove me to write, and what continues to drive me to talk about the Ghost horses, is the wild stallion I named Cuero, along with his herd. They became the signature horses of my study. Cuero trapped Hope and me with his gaze and drew me into the world of his fellow wildies. He maintains heroic prominence throughout my writing. I imagine him

thinking of me, his student: "She got it. It took her six years, but it was worth it." Of course, I also hope that I haven't intruded so much on Cuero's world that he thinks about me much at all.

◈ ◈ ◈

When I first read *The Man Who Listens to Horses*, I envied the author, Monty Roberts. He rode out with his horse, Brownie, and a packhorse to study wild herds in the high desert beyond the Sierra Nevada Mountains. I was captivated by his accounts of watching a mare school an errant foal. His studies of the wild Nevada herd laid the foundation for his interpretation of the language of equus, as well as his horse-training methodology. He writes that "binoculars were key" to his work, as he could watch every movement of the herd, untainted by human intervention.

The terrain Roberts worked in is very different from the dense mixed-boreal forest in which I quietly observed the Ghost horses. However, Roberts and I agree on our research methods.

I realized in 2009, while collecting video footage of the horses with a heat- and motion-activated video camera, that there were very few images to download. It was at this point that I figured out that the wild horses are mostly nocturnal creatures during the summer months. I wanted to know more about their habits without human presence, so I placed my first capture camera for still digital images near a mineral lick in February 2011. The capture camera is also called a "camera trap."

Hope, Cub and I, 2008.

Those images, and many more gleaned from the efforts of three more cameras spread throughout the wild-horse haven, turned the trick. A narrative evolved which revealed the wild horse as an animal forming alliances with other wild animals for survival. No matter how good one's hide, creep or long lens, wild animals know of a human's presence, usually sooner rather than later. Additionally, I know from sad experience how dangerous it is for wild animals to trust humans. Horse trappers have their corrals on the fringes of the area. Wolf hunters frequent the place. I easily imagine a wolf hunter setting bait with a dead wild horse. Looking further back in my own experience, I am still devastated to say that a poacher slaughtered my study bears in Kamchatka, a scenario I dread seeing repeated if any wild animal learns to trust me again.

It was obvious to me from the outset that I had discovered a population of wild animals that for the most part had not been messed with by the wildlife management practices of our time. Scientific study, however important, is intrusive. There are few areas left in Alberta where humans have kept their management paws at bay, more as a result of luck than intention.

My work has not been easy, but it has been rewarding. I have personally funded, through my art sales, all expenses for this six-year study. I did not receive a grant or any other form of financial assistance. I have been hiking or riding Hope into the Ghost Forest for six years in weather ranging from −30°C, with its biting chill, to 25°C, with its accompanying flies. Averaging a trek every two weeks for eight months of the year leads me to think I went on thirty-two-plus hikes or rides per year, so about two hundred visits over six years. Each trek in averages eight hours, bringing my study hours in the bush to 1,600 by the time I finished writing this book. Using four capture cameras, I have taken a total of about one hundred thousand photographs of the creatures in the Ghost Forest. As I worked through the summer days of 2012, I combed through about five thousand images per week.

As I rode through the area on my excursions, I did have many opportunities to see the horses. And so I have a combination of images taken with the capture cameras and with my own camera from the vantage point of Hope's back. A wonderful old horseman designed and built a sheepskin-lined camera bag for the back of my saddle that holds my Canon body and one long lens. With an attachment that doubles the capability of the lens stored in another pouch, I have four hundred magnifications – not a lot by today's standards but the limit for me. I have no time for fiddling with a tripod; if I am not ready for instant action, the shot is gone. Some of my fellow photographers blanch when I tell them my camera is hand-held. For the quick shots taken from Hope's back, I pull a smaller version, a Nikon P7000, out of my horn pouch and pray my horse stands still.

I invite few people to join me in my research – only those who volunteer to help and who take a blood oath of secrecy not to divulge the location of the wild herds are invited along. So why am I telling you so much now? The wildies need to be understood on levels beyond preconceived notions, with the knowledge that their future and that of their habitat, not only in Alberta but worldwide, is uncertain. Only then will their conservation be assured.

LIKELY CUERO'S SON (RIGHT) WITH THE BACHELORS. HE IS OUT OF THE CHESTNUT MARE THAT HAS BEEN IN HIS HERD SINCE I FIRST SAW THEM IN 2006.

CHAPTER TWO

Local Legends, Tales of Burning Love, Main Characters and Ghost Stories

The first time I came across anything related to wild horses was when I rode north of my land on the Ghost River into Crown land commonly known in 1972 as the Forest Reserve. I saw a stud pile, which is a pile of manure left by stallions. (At the time, I thought stud piles were a type of territorial boundary marking.) My next encounter with wildies was perhaps a bit more interesting: one day later in 1972, a wild stud jumped into my fenced pasture with the ease of a deer. It was gone as fast as it had arrived, retreating by flying back across the road and clearing a barbed-wire fence into the Rabbit Lake Indian Reserve, part of the Stoney Nakoda Nation's reserve lands.

I wondered about the wild stallion and the stud piles for many years, but as I was preoccupied with teaching at the Alberta College of Art and creating my first important bodies of artwork, I allowed thoughts of the wild to sink into the corners of my mind. At the same time, I realized the Valley of the Ghost was rich in history when I received a copy of *Big Hill Country: Cochrane and Area*. I loved the parts that shed light on Donna Butters's family, as it was on their land, a couple of kilometres north of mine, where I often went for a ride. It did not cross

my mind that the origin of the studs that deposited those mysterious piles was linked to the Butters' history and that of others in the Ghost River Valley. Their stories give life to the horses' history in the area as well as to their character. There are so many stories.

◇ ◇ ◇

Moving forward twenty years, in 1992 I was riding on the Butters Ranch grazing allotment (called the Aura Cache Permit) in the forestry reserve. This was one of my last rides up into the forestry reserve before travelling to the Russian Far East to study the coexistence of humans and grizzlies. I stopped to look at a rundown wild-horse catch corral. I felt pulled to dismount and walk over to take a closer look at the far side. At my feet lay a hunting knife in an old leather scabbard. It seemed a decent knife. I decided to liberate it from its wet, sandy burial site.

I later mentioned my find to Erik Butters, co-owner of the ranch and son of Donna, who laughed, saying, "That was Swampi's knife." I offered to give it to him. Declining, he told me the following story.

In the 1930s, Jacob Swampi, a Stoney Indian, respected

and befriended Erik's grandmother, Jean Johnson. Swampi lived in a basic cabin up Rabbit Creek on the Stoney Reserve, but he often stopped by the ranch house for a glass of milk or a meal. In his culture one did not knock. He often rode into the yard and gave this hoot-owl call that announced his presence. The Johnson family had no TV, refrigerator or telephone, and so nine times out of ten, Jean heard Swampi's call. But he did startle her a time or two when she came out of another room to find Swampi sitting at the kitchen table. That went on for a number of years. One dark winter night, Jean sat up in bed and woke her husband, Laurie, to say, "I hear Swampi's owl call." Sure enough, they heard it, and lantern in hand, found him lost up the valley on the far side of the creek. He was an old man by this time, and his mind was going. They got him into the house, gave him coffee and food. He died not long after that.

During his life, Swampi had borrowed the odd thing or two from the Johnsons and was very good at returning what he had borrowed. Once, however, he borrowed Jean's hunting knife in a sheath, and never came back. One day Jean asked him for her knife and, getting huffy, he said, "I do not have your knife!" She dropped her line of inquiry, as it wasn't that important.

A year or two after Swampi's passing, Jean again wakened in the night, saying she could hear Swampi's owl call. Laurie said, "Don't be silly, Jean, you know he is dead." The next morning, when they went out to do chores, her knife and sheath were sitting on the back steps.

Erik's story captured me. The story of the Ghost horses is also somehow a ghost story: you can see the ghosts of the past in the horses' bloodlines, in the way in which they are alert to their surroundings. I kept that knife. It lies in my kitchen cupboard below where I sat to write this book. The knife reminds me of the presence of the

Jean Johnson, circa 1930.

Jacob Swampi, circa 1930.

Stoney people and the Butters family in the Ghost River area, as well as their importance to the story of the wild horses there.

Butters Ranching Ltd. is a gorgeous spread about two kilometres north of my land along the Ghost River, stretching out on both sides of Highway 40. Erik Butters and his former wife, Wendy Fenton, took over the ranch from Donna and Richard Butters and developed it to what it is today. Erik and one of his two daughters, Erin, manage the ranch with Erin's husband, Darcy Scott. The grand matriarch of the Butters clan, Erik's mother, Donna, lives part time in the original log house. Richard died some years ago.

Erik told me a story about his father, who one day was riding in the forestry grazing allotment looking for a missing bull. He found it and also came across a wild filly, a week or two old, on her own. Richard thought she might have survived a cougar attack, as her haunches were scratched and one ear was chewed up. This foal stuck to his horse. Erik remembers seeing his dad chasing the found bull across the field and wondering what the heck was following him. "We had to feed the damn thing milk," Erik told me. Donna initially soaked bread in milk to get the filly to bucket-drink. She turned out a wonderful horse and an easy keeper. Erik loved to ride her in calving season during the coldest days of late winter. The horse hated a bit, so he didn't have to warm one up before riding her; he simply fitted her with a hackamore and rode out. Chico, as she became called, was an interesting horse. Erik says that if she came across a newborn calf, soaking wet, she would nicker to it and lick it.

The story of Chico and how she came to the ranch filled my imagination from the moment Erik told me the story, which was shortly after I moved to the Ghost River Valley. And her name became a touchstone for me later, during the storms that engulfed my small cabin in thick fog and rain when I was at Kambalnoye Lake in Kamchatka. I longed for the comforts of home, to be warm and dry. Huddled inside the cabin for weeks at a time, I often thought of Chico, the wild mare on the Butters ranch. There was something about Chico's spirit – after all, she survived a cougar attack and then elected to join a domestic herd – that reminded me of the three grizzly cubs Charlie and I were reintroducing to the wild. If the mare could become domesticated, I told myself, our cubs could survive in the wild. I named one of the cubs after her.

Chico was thirty years old when she died. For me she was a romantic icon, possessing a special gift she passed along to the domestic equines she encountered: her wild background.

Erik's daughter Erin said the wild horses figured large in the Butters girls' imagination. It was the biggest excitement of the day to see them when she started riding up in the grazing allotment when she was six or seven years old. Erin, with her beautiful smile and a soft lilt in her voice, described hearing the story of her favourite pony, Chico, following her granddad home. Erin says that she and her sister, Alyssa, proudly told people they had a wild mare.

Alyssa confided to me, "I loved her. Oh, man!" Alyssa believes that Chico's hatred of the bit was because she had

been born wild. She told me that when she rode Chico, she had to pay attention when they were near willows: the mare would take off through the willows as if they did not represent any kind of obstacle. There was nothing malicious about it, Alyssa said; she was just back home, was all. Alyssa holds a degree in veterinary medicine and is my vet. She and her husband live a few valleys over, but she remains a shareholder in Butters Ranch.

Erik's brother, Lamont Butters, works in Calgary but lives on his part of the ranch, saddling up his horse and skillfully lassoing calves at local brandings.

◈ ◈ ◈

Alyssa on Chico, the wild black pony; her sister, Erin, on her chestnut pony.

After the appalling slaughter of the Kamchatka study grizzlies, including my favourite of the orphan cubs, Biscuit, at the hands of poachers, I promised myself I would never again become involved in a project where animals' lives could be at risk because of attention I have brought to them.

Yet, in 2006, I began a study that would once again bring possible danger to a special group of wild animals. I accepted a challenge from my neighbour, Penny MacMillan, to find wild horses in Alberta. "Maureen, want to ride up on Butters grazing allotment? There are still wildies up there!" The Butters Ranch grazing allotment is on Crown land and was being considered for expanded ATV use by the Alberta government ministry that at the time was called Sustainable Resource Development (SRD). Penny and some of her friends were thinking about how they could fight the SRD proposal. I didn't want so much to get involved with Penny's battle against the government, but I did want to see a wild horse; so I accepted her invitation.

Penny owns land across the road from me and grazes a small remuda of horses, of which her favourite is the gelding Rocky, in spite of his fractious nature. I like to go trail riding with her, as she has no need to chatter endlessly. Penny understands my need to interpret tracks along the way as we ride. For some time she was barely able to see without glasses, which has led to some funny moments. I once asked her why she held her camera off to the side. "Well, I can't focus anyway, so I just shoot from the hip." On another occasion, she showed up at my place riding a restless colt instead of her gelding. When I asked why, she

replied in surprise, "What, isn't this Rocky?" Thankfully, two years ago she had eye surgery.

Penny is a bred-in-the-bone horse gal of the Ghost River. In 1884 her great-grandparents, Alexander and Mary Gillies, homesteaded two sections of land on the east bank of the Ghost River, at the confluence of the Bow River and the Ghost. The place has since been known as the Ghost River Ranch. It was sold in the mid-1930s to Agnes Hammond and Tilda Miller for about $10,000. My land was part of the Ghost River Ranch, as was Penny's.

When I rode out with Penny to find wildies that day, I asked her about when she first saw the wild horses. She said, "I was about fifteen years old, out riding with Donna and Richard Butters. It was amazing. Bays for sure, and that is all I remember. The ride was about gathering cattle, but we met some wild horses. I fell in love with it out there."

I told her of my first sightings of the horses in 1980, and how memories of those moments rest in a favourite corner of my mind. I was on a hiking trip just outside Banff Park boundary between Burnt Timber Creek and Clearwater River. I remember being awestruck by a herd of black and bay wild horses as they catapulted down a steep embankment, jumped into the river and charged up the other side. I vividly recall the musculature, the flashing of light through fountains of water, the intense speed – not a hoof was out of place as they raced over rock and through mud.

That day, under a bluebird sky, Penny rode down to my

corrals, loaded Rocky into my horse trailer next to Hope, and we were off to find ourselves some wildies. Having gained permission to do so, we drove across Butters Ranch. The Butters own 404 hectares, have a renewable ten-year lease on 809 hectares and pay, by the head, to graze an additional 8,093 hectares of Crown land, the Aura Cache Permit. Every year, the Butters round up two-hundred-plus head of cows with calves and move them up into the grazing allotment, which is also the almost unknown home of the Ghost wild horses.

Butters beef on the grazing allotment.

Driving through the allotment is a visual treat in mid-June, which was when Penny and I set out. Green everlasting and fat russet-brown cows and calves. There isn't a strand of barbed wire in sight. The Butters use a single strand of fine electric wire to divide their large holding to create smaller pastures of grass. They rotate their cattle from one area to another by simply rolling up the wire and creating another enclosed area, thereby avoiding overgrazing. Three-strand barbed-wire fencing is only used on the periphery of the ranch. Rolling grassland shimmers in pleasant contrast with vivid red-brown bovines grazing the lush grasses.

We branched north toward the Rabbit Lake Indian Reserve, and our passage onto First Nations land needed no signage. Bumping over a cattle guard, we noticed that the barbed-wire fence around the periphery of the ranch changes dramatically. On the ranch side, the wire is strung tight; on the reserve side, the wire lays in disarray, with some strands attached to fence posts and others in a tangle on the ground. A free-roaming bunch of bays, chestnuts and whites were in the distance beyond the lower marshes of Rabbit Lake.

Cody Wildman, a Stoney Native, lives near the marshland. We were about to cross the wetland on a good gravel road when a family of Canada geese waddled across in front of us and a red-winged blackbird darted from one bulrush stalk to another. We planned on asking Cody's permission, or that of his wife, to cross Stoney land. Erik Butters told us a sure way to remember Cody's wife's name: "I need a Wildman!" Yes, Anita is her name, and I got it right when we stopped to chat. I blew it, however, when I asked Cody if he was branded. He laughed, and I think I heard him say, "She wishes." I'm still not sure who laughed the loudest. I was referring to his horses, of course, but the gaffe helped Penny and me gain permission to continue on the road through the reserve. I must say that I have long appreciated the Native people I have known for their ability to laugh and enjoy the moment.

Driving on through the reserve, we emerged onto the old Owl Creek cattle trail, upgraded by Petro Canada (and currently owned by Direct Energy), and then we parked near one of the company's compressor stations below Salter Ridge in the Wildcat Hills, which, broadly speaking, are part of the Eastern Slopes of the Canadian Rockies. We saddled Hope and Rocky and headed up the old seismic line through a dense lodgepole pine forest. I remembered Donna Butters's direction, "Avoid the 'shot' holes." These are seven- to fifteen-centimetre-diameter holes drilled to place dynamite ten-plus metres down at regular intervals; these holes were once joined together by explosives wire, allowing the company to blow several holes at once to collect seismic data about the underlying strata. I did spot the odd yellow wire and hole, and I wondered if they would explode. Or was Donna warning me that a horse's foot could plunge into the holes – the latter, I hoped. I could have some control over that. With a little weight in my right stirrup, I guided Hope around them.

Pausing at the top of Salter Ridge to give the horses a breather, I took in, for the first time, the panorama of marshland, boreal forest and Rocky Mountains pressing their presence into the clear sky. That stop became a ritual for Hope and me in the six years to follow. The Devil's Head stands out prominently as a landmark known to all in this area of Alberta. I was excited to see stud piles appearing along the trail with greater frequency. Hope nosed down to smell a fresh deposit. Stallions smell stud piles and make fresh deposits to mark their presence in an area. A stud pile begins as a stallion stands guard over his herd, and after he and his group move on, it is thereafter added to by other studs.

Hope started to sniff every dwarf birch; her behaviour told me something was ahead. Suddenly she stopped, bringing Penny's gelding on a collision course with her rump. Hardly paying attention to the bump, Hope stared uphill, immobile. A white dot moved within the black shadow under the dense pine and spruce trees. A second white dot, half the size, moved slightly. A mare came partially into the sunlight, revealing a shining black body dappled with patches of sunlight filtering through needled branches overhead. The word "gobsmacked" comes to mind when I think about the experience now. A dark-bay stud let out a sudden blow and then, with his star-faced family, whirled, a flash of rippling muscle, flying tails and hooves. Gone! Only when the stud blew did Hope move.

She was gorgeous, with a fine-boned head, not the aquiline Roman-nosed misfit, the trash, that Donna Butters had once told me about when talking about the wildies. When riding the range checking on cattle during the 1950s, Donna likely saw hundreds of wild horses in any given summer. She was a superb rider who knew a good horse when she saw one, but she did have a bias: a horse wasn't a horse unless it was a thoroughbred. However, as I was writing this book, I asked her if she ever saw any good wild horseflesh on their grazing allotment. She reluctantly admitted to seeing the odd one.

After seeing the star-faced black mare with foal and stud, I was determined to find more wild horses in the Ghost Forest. Penny's invitation, and the Butters' stories, had me hooked.

◈ ◈ ◈

I had no idea where to start my search for other herds in the Ghost Forest. I called Erik Butters, sure that he would know where to find them. Being focused on finding stray cows, he hadn't paid too much attention to the whereabouts of the wild horses. He suggested asking the local horse trapper, who Erik insisted knew the range of every herd in the bush. At the same time, Erik doubted if the trapper, whom I respect in his wish to remain anonymous, would talk to me.

When I did call him, the trapper hesitated before speaking. He'd been expecting my call and finally agreed to see me, giving directions to his cabin on the edge of Rabbit Lake Indian Reserve. When I arrived, he offered me a cup of coffee. Canned milk and a jar of sugar were placed in the middle of the wooden table.

Early morning sunlight passed through somewhat murky-paned glass windows, illuminating part of the cabin's dark interior. I think he had electricity, but I couldn't see a switch for all the stuff hanging from pegs in the cabin walls. I saw horses on the hillside through the panes and kept an eye on their movements, thinking of my new 4 × 4 truck with its red paint, knowing it seemed to attract horses that liked to chew.

Before meeting the trapper, I asked a friend what he thought of his riding ability. "He is a cowboy who knows how to ride the forest and the muskeg – a good horseman – speaks a command and his horse jumps into the back of a pickup for him." My dad's horse, Perky, did that, too – no small feat. Another friend said, "They don't make cowboys like that anymore." I've also heard that the trapper loves his horses as much as others do their families. He earns part of his living as a cowhand and the rest from trapping horses and fur-bearing animals. He is referred to as one of the early entrepreneurs of the wild-horse world because early in the 1960s and into the 1970s he realized he could reap a cash crop from wild horses loose on Crown land. Like the Native peoples before him, he easily navigates the Ghost Forest on horse or by foot; he has a relationship with the wilderness born of both financial necessity and a love of living in the forest.

The trapper wasn't about to divulge where some of the herds were located, nor did he offer to guide me. He did offer to allow me to come and see a salt trap full of captured horses sometime, but he cautioned that it might get a little rough. Well, ranching is a rough industry at times, but nonetheless I declined. I just didn't have the stomach to see a wild animal trapped in a log corral. I remembered that he told a friend of mine that he chokes them down (lets the lasso tighten around the neck) when he catches them. At the same time, I have never heard of him being responsible for any atrocities conducted in wild-horse roundups or captures.

❖ ❖ ❖

Although I received no hints from the trapper on where to look, I decided to head out in search of the herds anyway. This trip marked my entry into the story of the Ghost horses, as I found in them a narrative rich with love, passion, jealousies and a variety of protagonists.

Topographic map, lunch and binoculars in my saddlebag, camera in my horn pouch, I rode out alone on Hope. She can backtrack her way over any kind of terrain without breaking stride, day or night. But she is hopeless in the muskeg and knows it. She does not have the muskeg experience of the wild Ghost equines, they having grown up running the wetland. She is flat-footed, unlike the wildies, many of which have teacup-shaped wild-horse hoofs. The teacup-shaped hoof is said have come from many origins – maybe from Spanish Mustang blood or perhaps simply the good feet passed down through the generations from an early escaped domestic horse. They certainly serve a positive function for running the marsh. I also think a wild horse pulls her hoof out of the muck immediately instead of letting her weight settle one foot at a time, as Hope does. It is always the upstroke that gets you, as I have discovered in rubber boots filling with ooze.

I headed up the seismic line, pausing for a breather at the stunning viewpoint on Salter Ridge, glassing for wild horses and paying my respects to the land. It didn't take long for Hope to walk out to this point automatically on our rides, as if to admire the view. This first day out, however, I did not linger, knowing the bugs would increase in ferocity as the day wore on, thereby making it unlikely that I would spot the horses. I thought I saw something red-brown out by a lake south of where the seismic line crosses the marsh, a thirty-minute ride down from our lookout point. I soon realized it was a deer, bounding across the marsh. I imagined the small sprays of water coming off the tiny hooves, flashing in the morning sun. Obviously something had startled it, so we headed down without delay. For my own purposes, I named that body of water Deer Lake.

We stopped near a spring emitting a strong smell of sulfur at the edge of the marsh not far from the lake; here, I scanned the wetland for movement. I could smell horse, too, possibly from a recent deposit on a stud pile. Or was there a herd nearby? Hope became immobile – this behaviour took me all summer to unravel – and swivelled her head downwind to a dandelion-covered meadow visible through the dense aspen grove.

Through the trees I spotted a group of horses that caught our scent downwind and immediately ran off. A bay mare was in the lead, and a small dark-bay stud with a very long mane and tail brought up the rear. Within minutes I felt the power of something staring at me. My eyes searched the dark shadows of the forest, and I saw a white blaze move ever so slightly. He was to the left of us, staring out from behind an aspen tree, a branch angling across his head and body, which further broke up his shape. He had obviously left the herd and circled back. He appeared to be hiding.

I couldn't sort out how the stud thought he was hidden from our view – patches of sun illuminated his hide. I wondered if he thought he could see us but we could not see him, certain of his camouflage. He stood motionless – except for the occasional swish of his tail – for about two minutes. Hope and I were caught in the power of his stare. I held my breath, returning his gaze, hypnotized. Finally my brain kicked in, releasing me from the trance. Pulling my camera out of my horn pouch, I rapidly took a few photos of the stud I later named Cuero, Spanish for *leather*. If he could survive with his band in this country, I reasoned, he was tough as rawhide, with some of the same rugged beauty.

He whirled and was gone. I recall starting to breathe again as Hope shifted her position with his departure. It was an epiphany kind of moment – something magical had occurred. I was over the moon as Hope pulled forward with her long-reaching stride, back up and over

Salter Ridge. The image of that small stud with the black mane to his knees, staring us down with dark eyes, haunted me. I fell in love with Cuero that day, and for the next six years he became my model, revealing, one step at a time, equus as wild animal.

I made a drawing of him immediately upon my return to my studio. It was the first drawing I completed of a Ghost wildy, and one I shall never sell. His long mane and forelock and one eye were almost all that was visible to Hope and me. He seemed caught in time with a branch of an aspen tree angled across his line of vision – splitting his world from mine, almost daring me to enter his. Cuero caught me with his stare and everything he represented: making a living off of Crown land. Later, the trapper told me he almost caught him once, but the stud evaded the lariat. I thought I saw a glimmer of admiration for the long-maned stallion in the trapper's eye.

I have experienced grizzlies circling around me to gather scent from another direction for identification – a manoeuvre often misinterpreted as stalking with a predator's intent. I have filmed deer circling to gather scent downwind, but when Cuero was watching us, he was upwind. He had already smelled us, identified horse and human, but he remained immobile behind that aspen, staring for two minutes. As I worked on my drawing of Cuero, I still felt the intensity of his gaze. At the time, I couldn't interpret his need to stare. I had more miles to ride and much to see before I could understand that stare. When

Cuero A Silent Study, charcoal drawing, 25 × 21 in., Enns, 2009.

I realized what was happening, I wondered why I at first found Cuero's behaviour so strange.

The following week, I rode Hope back to the edge of the marsh, stopping to admire the Devil's Head illuminated by the early morning sun. Seeing a band of horses out on the marsh, I quickly dismounted Hope, camera in hand.

Above: Cuero and his herd with the chestnut mare in foal.
Opposite: Cuero breeding the chestnut mare, 2006.

Romantic pair: stud finds his filly, 2012.

It was Cuero with his band. With no idea he was being watched and photographed, he rose on his hind legs and bred a chestnut mare. I have seen domestic studs rush at mares and breed them, often while the mare is hobbled, her tail bound, her parts washed. Like a machine, the stud mounts his objective; it's not pretty at all, rather coarse and violent. Here, the breeding was beautiful. Cuero's mane fell in a cascade, covering the mare's head and shoulder, a curtain that gave them private time. I hid with Hope behind some trees, feeling like a voyeur.

The following spring I photographed Cuero leading his herd of five across the wetland at the northeast end of Horse Lake – the chestnut mare last, heavily in foal. She has since raised two blond-tailed young stallions, now part of the bachelor herd. In 2012, with another foal at foot, it seems the blond mane and tail will carry forth to future generations.

Some outfitters told me of a time when a wild stud mated so aggressively with a wild filly that she galloped, with him in hot pursuit, into their dude horse band for protection. The pair quickly vanished in a flash of hooves and horseflesh, but the brutality of the breeding remained in their minds. In all my time watching the wild horses, I have not seen anything close to that. In fact, in 2012, I witnessed a courtship that would melt the most hardened of hearts.

A big bay bachelor left his herd in 2011, actively seeking a filly to start his herd. I saw him chased away from another stallion's herd in a thunder of hooves on frozen ground.

His flight continued up through the dense spruce and aspen forest to the crashing of dead branches and logs. The noise stopped. A few minutes later, the resident stallion emerged with ears back and head held low to the ground, snaking his way behind his waiting family, moving them across the marshland. Stallions weave their heads low to the ground when ushering their herds along, almost nipping the heels of the stragglers. It is like the motion of a cobra, body and head weaving with fluidity.

In June 2012 the same chastised stallion won his prize in an episode I would dearly love to have witnessed. She is one of the most beautiful fillies I have ever photographed. They nuzzled noses, twitchy lips and ears as part of their pre-mating ritual. I did not witness the mating but assumed it was beautiful. Women of our culture would love that bay stallion, as well as Cuero. The bay stallion for his romantic nature, and Cuero because he honours his lead mare, switching off lead and defence roles with her. When he leads, she is in rear defence position, as strong a mare as anyone could find. When she is in front, he steps with calmness and dignity, totally alert as rearguard.

Another stallion a modern woman would love is El Cid, another of the strong characters in the Ghost horse story.

◈ ◈ ◈

The old-growth spruce trees with roots anchored in the wet, moss-covered earth were illuminated by shafts of light, representing a scene reminiscent of an Emily Carr painting of cedars on British Columbia's coast. Carr is an icon

Threatened, mixed media photo painting, 22 × 28 in., Enns, 2010.

Into the Emily Carr Forest.

for me, as she was compelled to create art by the beauty of the forest and its connection to First Nations peoples. I wondered: If she were at my side that day, what would she have painted? Not wanting to be presumptuous, I painted my own version of a Carr series not long after, titling the four small works *If Emily Carr Painted the Ghost Forest.* Emily Carr (1871–1945) died a few years before I was born. I imagined seeing her forest flow into mine.

That day, I was awakened from my reverie and Carr and her cedars by the realization that no frogs were croaking in the marsh nearby. Birds illuminated momentarily in flight vanished with only the sound of moving air. There was silence in the swamp. Hope stood silently gazing in the direction my glance had fallen.

Into the Wild, painting, 22 × 16 in., Enns, 2010.

A tail swished from behind a tree set in the darkest of shadows, about fifteen metres away. Other than an occasional flick of his tail, the black wildy stood motionless. A bit deeper into the darkness of the understorey, I could vaguely see a second black horse. I would have missed them had it not been for finally paying attention to Hope's lack of motion and noting the direction of her gaze. I wondered how long they could stand there without moving. The pair had to know we were nearby. Hope too remained still.

After twenty minutes, my butt went numb and I fidgeted. Only at that hint of movement did the white snip on the nose on the otherwise blue-black body move. A body came into the light, and then a second one. The black stallion I came to recognize over the next five years by a jagged white line of hair between his nostrils – much like a scar – flicked his tail, wheeled and vanished into the darkness, his companion close behind. Air inhaled, the squishing of moss receded and it was silent once more. I wondered how long they would have remained there, almost invisible, if I had not wiggled in my saddle. They were both stallions and of the herd I had seen on one of my first trips in. At the time, I wondered what happened to the other bachelors.

I named the one with the white snip on his nose El Cid. He is one of my favourite stallions in the Ghost Forest, alongside Cuero. I spotted him first as a four-year-old leading a bachelor herd near Deer Lake. I like to give the Ghost stallions heroic names because I think they are

Above: El Cid, motionless for up to twenty minutes.
Opposite: El Cid stalks me.

tough leaders, surviving against all odds. I saw the film *El Cid* a long time ago and recalled the rugged leader, played by Charlton Heston, who fought off the North African invasion of Spain. Seemed a good fit for this stallion – strong neck, light flashing off his hide betraying the appearance of oiled black skin, emanating power and brawn.

I did not see El Cid and his companion again for two years. Then I spotted them across the willow-covered marsh at the north end of Horse Lake. I performed one of my best undetected approaches toward them in the thick willow. They had gained weight, bachelorhood clearly agreeing with their bodies but not their peace of mind. They were ogling a mare in another herd. El Cid has since proven to be a tough leader with a sense of mischief.

On another excursion into the forest in the spring of 2009, my partner, Rick, and I noticed a herd of horses on the north side of Horse Lake. On hands and knees, or a semi-crouch position, we crossed a hundred metres of frozen marsh, ducking behind bushes and hummocks to reach the far side undetected, or so we thought. Downwind and tangentially about fifty metres above the herd, we crept slowly from spruce tree to spruce tree. Above the herd now, Rick suggested I take my big lens and crawl forward to a tree that would put me above them in a central location. Perfect, I thought. Unbeknownst to me, Rick was rolling video footage of me on hands and knees, my departing ass full frame, aiming to hide behind the tree looming in front of me. I noticed that something was agitating the herd. A member was onto us.

"Shit!" I muttered. Moving quickly, I reached the tree. To my complete astonishment, there, peeking through the branches, was El Cid, curiosity written on his face, ears cocked forward, walking rapidly toward me. Rick later told me the stud had stayed hidden from me, using the same tree I intended to use. He knew I couldn't see him. The herd, milling about behind him must have heard my expletive. I was stunned by his proximity and cleverness. He completely outsmarted me.

El Cid recognized his quarry, and in a flash of flying ice, snow and clumps of frozen duff from under his hooves, he flung his proud black head aloft and galloped after his departing herd. Three dark-bay adults, a two-year-old and a yearling ran the frozen lakeshore and west along Aura Creek. I did not see El Cid again until three years later, in the winter of 2012, a sighting that was followed by several sightings in different parts of his summer range. He has a way of flinging his head to the right as he gallops away, one ear cocked to what is behind and one forward for what is to come. His long, full, black mane flies like a flag off to the side, whipping and snapping in the wind created by his passage.

During the very wet June of 2012 I hiked in to replenish batteries on the capture cameras and replace SD cards. Rick and my good friend Lois Green accompanied me. As we emerged from the marsh onto drier ground, they were ahead of me when suddenly I saw Rick gesture in the direction of the forest. We had crossed the first marsh, and the ground squelched with every footstep. Quickly

El Cid flings his mane as he runs off, every time.

adjusting my light levels for the darkness, I reached where Rick and Lois were standing behind a tree. Six years into the study, we were all well versed in the wild horse's manner of emulating deer by looking out from behind trees and thereby breaking up his or her shape. I detected the telltale white snip on a nose – El Cid's give-away – in the black shadows. He really has refined the

El Cid and his herd, Camp Ridge, 2012.

El Cid poses as if proud of his herd.

freezing-in-the-forest trick. I would have missed him if it were not for that white jagged line looking oddly out of place.

El Cid has a gorgeous lead mare and a small herd. We found him during 2012 on top of Camp Ridge, so named as it is easily glassed with binoculars from a place where we occasionally set up a tent. He seems to know me now, which is alarming, as I would rather he feared all humans. He seems to like to pose if caught in the open, clearly proud of his herd.

◈ ◈ ◈

Another of the Ghost's protagonists is a stallion Rick and I first saw when we were out gathering some footage and stills in early April 2009. After a long, circuitous hike through wet marsh, we found ourselves downwind from four wild horses. They were totally unaware of our presence. The big, stunning dark-bay stud stood beside an elegant bay mare that was at least sixteen hands high (one hand = ten centimetres) – this was Pocaterra and his mare, the Diva. His young son was not far off, and a mare in foal drifted closer from across the marsh.

Noticing us, Pocaterra ran in our direction, stopped to blow and was off again, seemingly elevated from the ice-covered marsh underfoot. His trot on springs of toned muscle and sinew sent him forward and upward to lightly come down, barely weighting his hoofs with every stride on the slippery footing. His mane flared in long waves, and his long, thick tail fanned out behind him like an Olympic banner. The Diva hesitated in making an opera singer's bow in the style of her namesake. The four exhibited shock and curiosity at our close proximity, huddled as we were behind the spruce with camera lenses clicking. They ran off, returned, circled, ice mixed with water cracking and flying, suddenly released from its winter state. Finally downwind, they had our scent. At full gallop they thundered across the frozen ground through the aspen forest, the odd branch cracking, making the sound of rifle shot. In a flash of muscled ass and tail, they were gone, and we heard hoofbeats disappearing on the wind. Wearing my artist's hat, I was overwhelmed by this stallion and his Diva's wild sculpted forms, and I completed two large drawings highlighting their individual splendour.

I named the big stud Pocaterra immediately, after George Pocaterra, who in 1941, with his opera-singing wife, Norma Piper, known as the "Diva," bought a small ranch along the Ghost River, almost two kilometres downstream from my land. Naturally I named his big, elegant bay mare the Diva. I love reading of George Pocaterra's fur-trapping days with his Native friend in the early 1900s in Kananaskis Country. Both George and his namesake took on any challenge, whether it was uncharted country in a mountain range or living with a Diva!

Someone once asked me why I name the studs of the herd and never the lead mares, with the exception of the Diva. This has become my practice because the studs are the ones who trot forward to investigate anything

Pocaterra, left, with young stud beside him; the Diva, with white blaze, behind black mare.

Pocaterra, charcoal drawing, 42 × 30 in., Enns, 2009.

The Diva Takes a Bow, charcoal drawing, 42 × 30 in., Enns, 2009.

Chief Goodstoney in sacred alignment.

potentially dangerous. I take tack-sharp pictures of these stallions if I am quick enough! The rest of the horses in the herds, lead mares included, are on the move by the time I swivel my attention. My style of photography does not allow for tripods. I quickly anchor my 400 mm lens on my body or against a tree if nearby.

◇ ◇ ◇

At one of the promontories jutting out into the marsh, with Hope tied to a tree behind me, almost asleep at the tree's base, I startled myself to attention seeing a horse's tail swish from behind an aspen across the marsh. I did not know at that moment that I would soon see a horse that would lead me further into the story of wild horses in Alberta as they relate to Native oral history and tradition.

With my always handy binoculars I made out the shape: one horse. I thought there had to be more, so I used Hope as a decoy and crouch-walked beside her as we crossed the marsh, being careful to step where the wild ones had crossed. A stallion suddenly spotted Hope and raised a proud head – backlit, silhouetted against the Rockies with the Devil's Head above his right shoulder – a mythical placement. He had a diagonal battle scar from eye to right nostril.

Not moving, his arrogant look full of well-earned pride gained by living free in his forest domain, reminded me of photos of the First Nations chiefs taken in the early 1900s. They posed with great dignity. I named that stud Caesar but later changed the name to Chief Goodstoney after

Chief Goodstoney Looking through the Aspen, charcoal drawing, 30 × 21 in., Enns, 2008.

Chief Jacob Goodstoney, the leader who signed Treaty 7 on behalf of his people. I imagined him with a self-reliant attitude, full of pride, almost arrogant as he looked over the foothills of central Alberta. I tried to evoke those feelings in my drawing of the stud now carrying his name. Both earned their outlook on the world. They were free, independent and a cut above the norm.

As the oral history goes, the Stoney First Nations people feared the mountain they called the Devil's Head, leaving offerings on its ledges for the spirits. Other oral accounts describe it as a site sacred to the Indians. The English place name is a translation from the Cree *we-ti-kwos-ti-kwan*; in Stoney, *si-ham-pa*. Peter Fidler was the first European explorer to refer to the peak by name and shape, in the late 1700s. Its 2,796 metres were first climbed in 1925, via the west ridge, by famous guide Edward Feuz Jr. and J.W.A. Hickson. Both of these men made many first ascents of major peaks in the Canadian Rockies.

The stallion Chief Goodstoney's pose that day was offset by the Devil's Head; in fact he and the mountain were positioned in sacred alignment, with the mountain over his right shoulder. I have not seen that stud since that moment. The drawing I completed of him is haunting, with eyes that follow anyone passing. A friend who visited me once felt the energy of the drawn stallion's stare.

In an informal discussion with Dr. Brian O.K. Reeves, a professor emeritus of archaeology at University of Calgary, I asked if a wild horse could be a sacred site. He kind of smiled and said that applying that word to a horse didn't fit. I also talked to Peter Snow, a Stoney environmental consultant I met when searching for reasons to save the southeast corner of the Ghost Forest, my study area. I asked Peter if there might be sacred sites within the area. Peter said a study needed to be done but told me I had already found a sacred site, as the wild horses themselves were sacred to the Stoney people.

When I went back to Dr. Reeves with Peter's response, he told me that Snow likely meant "revered." Dr. Reeves defines sacred sites as "places of all kinds, including the built environment, where visionary experiences, including miracles of healing, occur. To qualify as such, these transcendent experiences must be experienced by more than one person, be transgenerational in nature and validated by the community. For example, Chief Mountain."

My friend Eleanor O'Hanlon wrote me in response to my question about sacred sites, quoting the Lakota holy man Black Elk, who around 1947 said, "We regard all created beings as sacred and important, for everything has a *wochangi*, or influence, which can be given to us, through which we may gain a little more understanding, if we are attentive."

◇ ◇ ◇

During my first year finding the story of the wild horses, I found my subjects more interesting than grizzly bears, as they seemed to hold more behavioural surprises. Nothing has been definitively written about how they became wild animals, easily dismissed as "feral" or free-roaming escapees, nuisances on the range like rabbits in Australia. Because of preconceptions of these horses as merely feral, they have never been the focus of serious research; they have never been allowed to tell their story.

Despite my interest in them, I was going to quit during the winter of 2007, thinking it was useless to fight a mindset entrenched for decades in the thinking of government and ranchers. I was doubly certain I could not deal with another slaughter, this time in the Ghost Forest. The death of the bears in Kamchatka was still raw. But I was drawn back into the southeast corner of the forest of the Ghost River Valley by forces unknown. The presence of the horses took hold of my mind and imagination – vivid images, memories, haunting art and dreams. Maybe it was the ghosts of the area not willing to let me go, pulling me back to their realm.

Roland Gissing (1895–1967) made his home for many years at the junction of the Ghost and Bow Rivers, downstream from where I have my studio and land. Coincidentally, his first wife was the great-aunt of my horseback-riding buddy Penny, who is part of this evolving story. In Max Foran and Nonie Houlton's book *Roland Gissing: The Peoples' Painter*, the artist is quoted as saying: "The whole of the country surrounding the Ghost River, even to where it rises, was connected with ghosts and spirits … I myself can remember, many years ago, seeing Stoney Indians lashing their horses to a gallop to get across the river before sundown if it happened to get late and the sun was setting."

On Captain John Palliser's 1860 map, the Ghost River is shown as Dead Man River. My friend Penny's great-uncle John Gillies told a story to Gordon Hall, a reporter for the *Cochrane This Week* newspaper, which was published November 7, 1989:

A white horse came down the centre of the river. Astride its back riding backwards was an Indian warrior. Roland Gissing painted this scene. Another one was of a hunting party returning from up the Ghost. When they came to the top of a hill they could see their lodges ahead, cooking fires were visible with people moving about. But when the hunting party arrived at the camp, there were only dead bodies there, the camp was silent and the fires were out. Smallpox wiped them out. The late John Gillies spoke of ploughing up skulls and reburying them at the confluence of the Ghost and the Bow River.

⇻ KIT'S BAY FILLY SHOWS EVIDENCE OF GOOD BLOODLINES. ⇺

CHAPTER THREE

Wild-horse Bloodlines: Stories and Histories Imperative to Alberta's Heritage

Recalling the late Andy Russell's spellbinding stories, I would love to have been able to ask him for his best one linking Alberta's wild horses to Alberta's heritage. Every wild-horse lover I have talked to mentions their historical value, thinking surely that consideration of their history will stop wild-horse culling permits being issued by Alberta's environmental governing agency, now called Environment and Sustainable Resource Development (ESRD). But up until now, no one has defined the historical value of wild horses beyond that of Spanish Mustangs.

While mustangs do have some relevance when considering Alberta's wildies, there is more to them than that: they do have interesting bloodlines. By the early 1900s, there was a mix of Aboriginal and non-Aboriginal horse genetics running through Alberta's wild horses. How did these beautiful athletes of the Ghost Forest gain the genes they need to survive in such perilous conditions?

To answer that question, I first accepted that all of Alberta's wildies have domestic ancestors that were introduced to a wild landscape. I also understand that their history both predates and encompasses European arrival in North America. As I set out to discover more, I was interested in hearing the oral stories of the Piikani Nation about the first arrival of horses to Alberta, as well as the oral accounts of the Stoney Nation as to how they originally came by their mounts. The Stoney people's lands, of course, border the Ghost Forest study area. I compared the oral narratives to the written one of David Thompson, who spent many winter nights in his teepee in 1787–88, listening to Chief Saukamapee's stories of how the Stoneys took horses instead of scalps from the Snake Indians.

The oral stories and those of Thompson corroborated the Spanish Mustang / Andalusian phenotype evident in some of the horses but not the draft horse and fine-boned thoroughbred types. That influence came from somewhere else. Asking about this, I heard more than once that the Ghost and Grand Valleys were known as the valleys of good horses in the early 1900s. These valleys are about twenty kilometres apart, but there were no fences in those days. Ranchers' ranges overlapped what is now the Ghost Forest study area, bumping up against Stoney lands too.

With no roads, only wagon trails, the early ranchers and homesteaders needed draft horses to work their fields and mounts to ride in order to find their cattle. These pioneers were as tough as their horses – both had the will to survive. But the early settlers of these valleys were also attached to the Old Country, and they imported thoroughbreds with which to play polo. Those olden-day ranchers and their horses would eventually form a link with Alberta's wildies, making the Ghost River Valley horses integral to Alberta's heritage. But first, the ancient history.

◇ ◇ ◇

When Rick and I visited the Pech Merle cave in southern France in 2011, I was stunned by the sophisticated drawings of the spotted horses, carbon-dated to twenty-five thousand years ago. Clearly, in Europe the horse was one of the necessities for survival, to have been revered and depicted with such clarity and sophistication. I wondered how that came about, and how the horse came to North America. Surely there was more to the story than what I knew about Columbus arriving in 1492, presumably with some Spanish horses.

Fifty or sixty million years ago, the first horse-like creature, *Eohippus* ("dawn horse"), lived in both North America and Europe. It was a small animal, barely a foot tall, that became extinct in Europe but continued to develop in North America until it finally went extinct for tens of thousands of years. Some apparently survived, however, having crossed the land bridge into Asia, and these developed into the horse similar to what we have today, *Equus caballus*. According to scientific dating, some returned to North America about six hundred thousand

Horse in the Pech Merle cave, drawn on stone twenty-five thousand years ago.

years ago, and they remained here until about twelve thousand years ago, when they became extinct in North America but continued to exist in Europe and Asia.

According to anthropologist Dr. Brian Reeves, radio-carbon-dated horse remains were found in Wyoming that place horses there before the smallpox epidemic of 1730. In the early 1960s, horse bones were also found at the Cluny Fortified Village, on the north side of the Bow River in south-central Alberta, radiocarbon-dated to 1730.

A historical review of early fur trade explorers' records of and interviews with the Xeni Gwet'in (Nemiah) First

Goodstoney, mixed media drawing, 22 × 15 in., Enns, 2006. I visited Argentina's Salta area in 2005 and was impressed by primitive Inca drawings in the caves in an area accessible only by horse. I thought of the stallion in front of the Devil's Head.

Nation in British Columbia also indicate that horses were in the Chilcotin–Nemiah area before European contact. When I found out about this, I did not know how to reconcile ancient Aboriginal beliefs, passed on through the generations, with scientific fact. So I talked to my friend Eleanor O'Hanlon, who was conducting research about ancient understandings of animals as guides to self-knowledge for her new book, *Eyes of the Wild*. She said I should try to keep in mind that when First Nations people say that horses were present before the arrival of Europeans, they are speaking a spiritual truth. As I understand it, the spiritual essence in all creatures is not bounded by linear time. I think that for First Nations peoples, it is their oral history that holds value, not a carbon-dated bone!

Hank Snow was a member of the Stoney Nakoda Nation band council when I was working on an educational project about wild horses in 2009 called *Wild and Free*. I interviewed him for the project, and he laughingly told me that he found it funny that white people always seem to claim credit for the arrival of the horse in North America. There are many accounts in Stoney oral history of the first horses coming from the bottoms of lakes. On Hank Snow's invitation, Dandy Amos, a fifth-generation Stoney, spoke and was recorded for the project's website. He talked about catching wolf and coyote pups and raising them for packing and pulling the travois. He added an account of an old man dreaming, who recalled he could catch a horse from the deep waters in Kananaskis Lake.

❖ ❖ ❖

On his second voyage to the new world (1493–96), Christopher Columbus brought at least twenty-four horses, a mixture of Barb (North Africa desert blood) and Andalusian (Spanish blood) to the Spanish horse-breeding stations on the Caribbean islands of Hispaniola and Cuba. Within a few decades, Spain controlled Mexico, Central America and parts of South America. More and more horses were imported, and by the time Francisco Coronado led his expedition north across the Rio Grande in 1539, he had two hundred horses with him. Large numbers of ranchers settled in what is now New Mexico, and horses were vital for work and prestige.

The Spanish masters of the horse correctly believed that allowing the Indians access to horses would empower them. In 1621 the law forbidding Indians from riding horses was lifted, as the ranchers needed mounted herders. Some horses were stolen and others strayed. North American Indian vaqueros, horse-mounted livestock herders more or less enslaved by the Spanish, soon learned that owning a horse was the next-best thing to freedom. Before long, bands of wild horses ran free across North America. Wild herds grew and expanded their territory over the next 150 years, reaching Canada by the mid-eighteenth century. "Mustang," the name for the horses that ran wild, comes from *masteno*, referring to a type of stray of the Spanish stock raisers.

No one knows for certain how many wild horses roamed the West at their peak, but by the mid-nineteenth century, conservative guesses put the numbers between three and five million. At the same time, ranchers' cattle roamed thousands of acres. With the enactment of the Homestead Act, settlers claimed the prime land, where there was water. The settlers hated the cattle barons' cattle because they gorged on their crops. Barbed wire was invented in about 1870, bringing an end to the land of the wide-open range. Fifty years later, the numbers of wild horses was reduced to hundreds of thousands in the American West. By this time, the mustangs had mixed with other introduced European horses, such as thoroughbreds and draft horses. Ranchers began capturing them, hoping for a quick dollar.

◈ ◈ ◈

A German film crew made a documentary several years ago on the wild horses of the Nemiah Wild Horse Preserve in British Columbia's Chilcotin. They were intent on knowing how the original mustang came to run in the wilds of Canada. Mustang blood is known internationally, and somehow the relationship between Canadian wild horses and mustangs gives the horses value. In fact, DNA analysis of the horses in the Chilcotin, and discovery of mustang genes, helped to create the Chilcotin Wild Horse Sanctuary / Brittany Triangle. The Pryor Mountains in the United States have also been shown to have populations of genetically proven Spanish Mustangs.

That film crew and many others think of wild horses only as Spanish Mustangs – they cannot imagine that any other lineage could possibly constitute wild-horse status.

The mustang is a breed with identifiable DNA and a history admired the world over; however, it is not the definition of a wild horse. I visited an organization in New Mexico five years ago that claimed to be protecting wild horses by keeping a herd of Spanish Mustangs running free in a couple of fenced-in quarter sections. What they were in fact doing was raising a breed of horse, which reminded me of Barry Oakman, a man I met in New South Wales, Australia, who claimed a similar use of the word "wild" for the dingoes he raised in captivity for breeding. Barry Oakman is a great guy, and well meaning, much like the folks in New Mexico are well meaning in their intention to release horses into the wild that carry the DNA of once-wild animals. However, such DNA-focused projects do a disservice to the Ghost horses, which are just as in need of protection and study as a pure-blooded mustang herd, and whose other bloodlines are just as important to their genetic makeup and ability to survive in the wild. I do believe that mustang genes would be found in the Ghost horses – I think Cuero shows evidence of Andalusian/mustang blood in his physical type and carriage – however the horses' value in my study area is in their behaviour and their mixed blood, which in combination allows them to survive independent of humans, as wild animals.

Testing has been done for Spanish Mustang blood in the wild herds near Sundre, Alberta, and some horses carry the bloodline, but it has been significantly diluted. I asked Donna Butters what she thought of the possibility of mustang blood in the horses that run on their grazing

Cuero shows some mustang blood in this charcoal drawing titled *Cuero's Last Stand*, 40 × 30 in., Enns, 2008.

allotment. "Bullshit," or something like that, was her response. And really, whether or not the mustang genes live on in the Ghost River Valley is inconsequential to how important the area's wild horses are to Alberta's heritage, as the oral histories of the First Nations tell.

◈ ◈ ◈

When Anita Crowshoe visited my studio, she explained that the Stoney Nakodas have their creation story of horses, and she respects it. She went on to say that the Stoney Nation accounts of horses coming from the depths of lake waters is theirs to tell. First Nations peoples do not believe in the concept of dominion in storytelling, where one version is better than another. Her horse story is different, and it comes from a more recent time, having taken place in the 1700s.

Anita told me about horses coming up into Canada with the Blackfoot Nation, of which the Stoney Nation was considered an invited family member. The Stoneys came with their horses from North and South Dakota (thus the name Stoney Nakoda Nation). The Piikani (once called Peigan), one nation of the Blackfoot Confederacy, suggested they move onto land in the north near the mountains, to country with which they were more familiar. The Stoneys, along with their horses, continued north to the area along the Bow River and farther north to the Red Deer along the front ranges of what is now known as the Canadian Rockies.

After hearing Anita's story, I read the Chief Saukamapee story as recorded by David Thompson in 1787–88, for his account of the horses coming to the Blackfoot and to the Stoneys. I found Thompson's story difficult to grasp but reminded myself that it was told to David Thompson by Saukamapee (Young Man), who spent the winter of 1787–88 with Thompson. During this time, the chief, over many evenings in his teepee, told of the coming of horses to the Blackfoot in around 1730. Chief Saukamapee was between seventy-five and eighty years of age at the time.

The Piikani were the tribe that existed on the edge of Cree territory and upon whom the Snake Indians made their attacks. The Piikani came to Saukamapee's Cree tribe for help when Saukamapee was about sixteen. His father agreed to help and bring some of the Crees with him. They had a few guns; the majority of their weapons were lances with stone or iron points, larch bows that came up to the chin, and quivers full of arrows. Saukamapee himself had a bow and arrows, as well as a knife. Twenty Cree warriors joined up with about 350 Piikani warriors. Saukamapee remembered that the Snakes outnumbered them and were better armed. The iron-headed arrows of the Piikani and Crees did not go through the Snake shields but merely stuck into them. On both sides, several were wounded, but none died and the battle ended.

Many years passed and Saukamapee became a man. Much had changed when the enemy Snakes came to fight again; this time, they brought allies. The allies of the Snakes were the Big Dogs, the Cree name for the horses on which they rode, swift as deer. Saukamapee and his people had not seen horses before. They persevered and fought alongside their Stoney allies, taking scalps as war prizes. The Snakes were eventually defeated because they did not have the same gun power as their opponents. The Stoneys fought the Snakes in more battles with Chief Saukamapee, who began to advise the warriors to bring back horses instead of scalps, which they obviously did.

After reading Thompson's narrative about Saukamapee, the Stoneys, Snakes and horses, I asked archaeologist Dr. Brian Reeves about horses coming north with the Stoneys when they were friends with the Blackfoot Nation. He suggested that the horses came from the north, not the south, to the Stoneys.

"Stoneys were recent arrivals," he said. "Chief Goodstoney came south from the North Saskatchewan River in early 1800s. Two bands arrived later, crossing westward from the Assiniboine reservation, in northern Manitoba. Stoneys were not part of the Blackfoot Nation or the Blackfoot Confederacy, which included the Tsuu T'ina and the Gros Ventre. Stoneys probably stole their horses from the Blackfoot tribes in the late 1700s while still resident on the North Saskatchewan River."

I realize that the oral history accounts of the coming of the horses to the Stoneys and the scientific record differ somewhat, but I've concluded that there really is no right or wrong: there are only differences within oral histories and written documents and research.

◇ ◇ ◇

In his book *These Mountains Are Our Sacred Places: The Story of the Stoney People*, Chief John Snow also speaks of the taking of scalps and later horses as part of the spoils of battle. According to Snow, Stoney hunting territory at first went as far north up the Rockies' corridor as the Brazeau River and as far south as Chief Mountain. By the mid-1800s, the Stoney population was about two thousand, and their lands extended from the headwaters of the North Saskatchewan to the US border. The horse had replaced the dog, and firearms the bow and arrow. There were likely more horses than Stoneys, all of which ranged free near the settlements or were completely wild. However, with the signing of Treaty 7 in 1877 came the surrender of the Stoney lands and the agreement to move onto reserves with their horses. Today, the Stoney Nakoda Nation is made up of the Bearspaw, the Chiniki and Wesley First Nations. The largest group of Stoneys now lives at or near Morley, Alberta, with a small community on the Rabbit Lake Indian Reserve under the Stoney tribal administration at Morley. Without a doubt, when the Stoney people moved onto settlements, many horses eluded the transfer, continuing to run free.

By 1910–30, Stoney horses roamed wild north and west of the Morley Reserve and could easily make their way over to the Ghost Forest. There were no fences and only a dirt track where Highway 40 is located today. Additionally, with the Rabbit Lake Reserve on the Ghost Forest area's eastern boundary, more Stoney horses drifted in and mixed with whatever European stock was loose.

Stoney horses are for the most part free-roaming on the reserve, although some families at Morley have contained pastures with more-controlled breeding programs. Some

Spanish Mustang blood may now mix with the thorough-bred, Appaloosa, quarter horse and draft (Desjardine Percheron) lines evident in Stoney horses. I have heard from more than one source that to breed-up their horses, Stoney people capture wild mares and cross them with their domestics. The Stoney horses are mostly blacks, chestnuts and bays, although there are some examples of the respected white horse of their oral history pre-European contact. None of the white, grey, pinto or Appaloosa colorations are seen in the wild Ghost herds, however, which makes good sense.

Over several generations of breeding in the Ghost Forest, likely due to natural selection, the blacks and bays have survived whereas lighter coloured horses have been culled out – a light grey or a roan, for example, is too easily spotted by predators, human or beast. Quite the opposite coloration survives amongst the wild horses in Nevada, where duns, greys and light chestnuts blend into endless sand and rock.

Some mixed colours appear in the ferals and free-roaming domesticated horses using the Butters grazing lease. The odd one carries the brand of the local trapper. On more than one occasion when I've seen Anita Wildman on the road through the reserve, she has asked me about a missing horse. The Wildmans' horses are free-roaming domesticated horses, as are those of the trapper, and by behaviour alone they are easy to pick out, lacking the alertness necessary for survival in the wild, in addition to being of a colour that's out of place.

It is pure speculation on my part, but I see the influence of Native stock in a palomino and bay I photographed near Horse Lake. The palomino looked out of place; perhaps he was a recent escapee and not necessarily from a Stoney stable. Escapees to the wild come from non-Aboriginal herds as well.

◈ ◈ ◈

The Ghost River and Grand Valley, west and north of Calgary, were known in the early to mid-1900s as the Valleys of the Great Horses, thanks to the Kerfoot and Johnson families, and others who were interested in fine horseflesh. These families raised thoroughbreds for polo, many of which escaped to run wild. Having noticed the thoroughbred phenotype in some of the Ghost horses I have seen, I wondered if there were any stories related to how thoroughbreds may have leaked into the wild population. I was fortunate to track down a few possible links.

Hamish Kerfoot, whom I spoke to in mid-July 2012, proudly showed me pictures of his father and grandfather with their finely bred imported polo ponies. Hamish's stature fairly grew as he passed me the papers of his great-grandfather's stakes-winning stallion Porten and his grandfather's equally storied stallion Vambrace. The Kerfoot horses carried the Crown brand, known to all in western Canada and into the US.

Hamish drove in from Providence Ranch in the Grand Valley to a café in Cochrane, where we spoke of the

The Trapper's free-roaming horses on the Butters grazing allotment, 2007.

much-admired Kerfoot horses. Hamish still lives in the heritage house his grandfather, Duncan Irving Kerfoot, built with his English bride in 1912. I have known Hamish since I moved to the Ghost Valley. He doesn't mince words and says what he thinks and feels. His comment about his age made me laugh, but it is typical of his forthright way of speaking. He said, "I am fifty-four and don't give a rat's ass!"

Palomino stranger with bachelor at Horse Lake, 2009. I did not see this colt again.

He told me about a time he was in a bookstore in Cochrane, where he met an old cowboy with bowed legs – Kewley was his name. The cowboy said Hamish had to be a Kerfoot, as he looked like one. He told Hamish that in the 1940s he was grubstaking it and staying at the Johnsons' and other places. Kewley told Hamish how he and his family used to go hunting for horses in the Forestry lands and caught wild horses. They especially wanted to catch the ones with the Crown brand. Those were considered the prize.

I asked Hamish how Crown-branded horses could have been found amongst the wildies. "Did a bunch of thoroughbreds escape? If so, where did they come from in the first place?"

He told me it all started with his great-grandfather, W.D. Kerfoot, who brought sheep up to the Cochrane Ranche

Drawing by Hamish Kerfoot of his two variations on his family's Crown horse brand.

in 1883 – crossing the Bow River at Calgary with eight thousand head in prime condition. To make a long story of family history short, W.D., never too keen on sheep, eventually bought land to ranch in the Grand Valley. He imported Porten from England at a time when only ocean liners transported livestock. I was amazed at the amount of work and cost it must have taken to bring the horse to Alberta, and I wondered why W.D. would have bothered. But W.D. was a polo player and wanted to breed thoroughbred horses. Unfortunately, not long after Porten was brought over, W.D. was killed during a Calgary Stampede parade when a horse spooked, reared and fell backward over a cow, breaking W.D.'s back.

Hamish's grandfather, Duncan, born about 1885, took up the mantle of his father. On July 9, 1925, Duncan purchased what would be another fabulous Kerfoot stallion – Vambrace, bred by Lord Rosebery in England in 1920. By this time, the family was selling railway cars full of polo ponies to horse lovers in California and the eastern US. During a ten-year period they made more money on horses than they ever did on cows.

After the stock market crash in 1929 and the ensuing Depression, the market for their fine thoroughbred ponies dried up. At this time, the Hanson and Anderson families a few miles away along Beaupre Creek had a mink and fox farm and took all the old horses. (At this point in Hamish's narrative I remembered a term I'd heard during my childhood: "fox them," which means sell the horses for fox meat.) When the Depression hit, though, even the

W.D. Kerfoot in the early 1900s with his imported thoroughbred stallion Porten.

Kit – a stallion with some breeding

fox- and mink-fur markets crashed. As a result, the fur farmers could no longer take on their neighbours' surplus horses.

There were log corrals, rail ones and likely some barbed-wire fences in the Grand Valley at this time. Most working ranch horses and breeding stock were probably kept close to the main ranch buildings, but out of necessity, many ranchers let their horses roam free. I asked Hamish how many escaped the roundups or were turned loose when the horse market bottomed out. He said, "Could easily have been a hundred head; that wouldn't have been impossible. We had lots of horses. They didn't escape. That is like the trickle-down theory. Money doesn't trickle down unless there is a damned leak! They weren't worth anything." In other words, Duncan and others opened wide their gates, didn't round up those that were roaming in the forest, and their horses escaped. The strongest ones lived on to breed wild and free.

Wildies here show evidence of thoroughbred bloodlines.

Wild stud races forward to defend his herd.

When Hamish was growing up, his family often visited the Johnsons at the Lazy JL on the Ghost River, which was later called Butters Ranch. The Kerfoots and the Johnsons shared a passion for the galloping game, polo. I was keen to talk to Erik Butters about his grandparents, the Johnsons, and figure out how they fitted into my narrative about escaping horses adding bloodlines to the wildy herds.

I drove up to the Butters ranch on a gorgeous sunny afternoon in late June to talk to Erik. Erik had hewed pine logs from his woodlot with his former wife, Wendy Fenton, to make barns for horses. After I had passed one of those heirloom log barns on the way up to the house, I parked my truck to enjoy the rolling acres of green that only stopped when I got in my truck again and drove to reach the paved road of Highway 40, a couple of kilometres away. Log fencing built in the historic stacked style separated clipped grass from lush cattle browse.

Erik, born in 1952 and the eldest son of Richard and Donna Butters, was chairman of Alberta Beef Producers and is respected across Canada as a conservation-conscious, progressive rancher raising free-range beef. He is the forerunner in supporting riparian health of even the smallest waterways within his grazing allotment.

Erik showed me binders of family photos spread out on the coffee table, indexed in his handwriting. This man loves history. As he turned page after page of the old-style print photo album, he spoke with rapid enthusiasm. We chatted, and I began to realize that Erik spoke about his amazing grandparents and their horses as if they were a single family unit – and they were. The photographs of Indian chiefs, lots of people on horses, log cabins and families posing conjured up for me romantic images of the dawn of the twentieth century. However, judging by Erik's account, his family's homesteading life was anything but idyllic.

Erik's great-grandfather was Everett Johnson, who came to the Bar U Ranch in southern Alberta in 1889 and whose life as a cowboy and horseman formed the storyline for the famous 1902 book by Owen Wister called *The Virginian*.

The book begins with a tale of a cow pony evading a lariat as a train pulls into the station. One of the book's characters, a passenger in the train, tells of what he saw, with great admiration:

> The pony in the corral was wise, and rapid of limb. Have you seen a skilful boxer watch his antagonist with a quiet, incessant eye? Such an eye as this did the pony keep upon whatever man took the rope…. [T]he rope would sail out at him, but he was already elsewhere; and if horses laugh, gayety must have abounded in that corral. Sometimes the pony took a turn alone; next he had slid in a flash among his brothers, and the whole of them like a school of playful fish whipped round the corral…. Then for the first time I noticed a man who

sat on the high gate of the corral looking on. For he now climbed down with the undulations of a tiger, smooth and easy.... I did not see his arm lift or move. He appeared to hold the rope down low, by his leg. But like a sudden snake I saw the noose go out its length and fall true; and the thing was done. As the captured pony walked in with a sweet, church door expression, our train moved slowly on to the station.

Erik's mother, Donna, was a horsewoman from the day she was first lifted up on a horse as a young girl, and she was guiding for the Trail Riders of the Canadian Rockies by the time she was thirteen. Her granddaughter Alyssa carries an inherited ability to form partnerships with horses – the same trait that her great-great-grandfather Everett "the Virginian" had.

I saw evidence of that distant ancestor's abilities when Alyssa came to my ranch to give the fractious horse Amigo a fine syringe of strangles vaccine up the nose. Prior to her arrival, I had tried for hours to administer the drug, only managing to sweat up the horse and myself. Alyssa quietly walked in and talked Amigo into accepting the tube up his nose. She repeated her magic time and again, once when investigating lameness in Amigo's left foreleg. She didn't get kicked while she injected a local anaesthetic into his tendon; instead she lulled him into a state of calm with her sweet voice. I truly think that her predisposition for calming rebellious ponies is similar to some horses'

Donna Butters, guide, CPR trail ride to Mount Assiniboine, 1943.

Jean Johnson and her packhorse. Johnson homesteaded and was a tough pioneer woman in her own right.

A view across the marsh to the Devil's Head likely similar to what Jean Johnson would have seen.

ability to return to a wild state; it's in the DNA, passed on from one generation to the next.

The foreman of the Bar U sent for Harry Longbough (later known as the Sundance Kid) to come up and "break" horses, as Everett Johnson had recommended him. The Sundance Kid was best man at Everett's wedding; his signature is on the wedding certificate. As the great story goes, when ranching got tough, good cowboys like Harry Longbough went on to more lucrative careers, and robbing trains was one of them. Butch Cassidy (Leroy Parker), by 1890 well known as the Sundance Kid's partner, did not come up to Canada. I asked Erik if the Sundance Kid had ever holed up at the cowboy's hangout near Beaupre

Creek, called Robbers Roost. He did not know about that. But I would like to think so, given my romantic outlook on life. No doubt the Sundance Kid and his cowboy friends also rounded up free-running horses for an easy dollar. Robbers Roost is still around, about three kilometres south of my land.

Laurie and Jean Johnson were Erik's grandparents. They paved the way for a line of settlers and early guides of the CPR Trail Riders in the Rockies. They lived with few

Polo ponies and their riders, 1925: (left to right) Laurie Johnson, and Archie, Pat and Duncan Kerfoot.

dollars and no luxuries during their early ranching years. Jean homesteaded some land on Rabbit Creek, called Jean's Creek, in 1932, while Laurie homesteaded more land in 1936, which was later known as the Lazy JL Ranch. Jean Johnson was as strong a female pioneer as was ever known to live in Canada. In the early 1950s, she camped alone in a tent ten days at a time, tending cattle, riding back to home with her packhorse only when she needed supplies. Laurie was a horseman par excellence. Their daughter, Donna, married Richard Butters (Erik's dad), who owned the OC Ranch along the Ghost River bordering the Lazy JL, formerly owned by Jean and Laurie, thereby amalgamating the holdings and creating the large ranch in the Ghost River Valley that it is today.

Erik and Hamish's stories join when they speak of their grandfathers, who both were serious polo players and lovers of fine horses. The Grand Valley polo practice ground was the front field of a spread now owned by Simpson Ranching. It was formerly part of the Kerfoot Ranch.

A photo of Laurie Johnson mounted to play polo caught my eye when I spoke with Hamish and looked at his collection of photographs. Laurie played the galloping game on a fine specimen of a horse. I had a hunch, so I showed Donna Butters the papers of Duncan Kerfoot's horse, Vambrace. Donna lit up like a Christmas tree. "Vambrace! He sired my father's thoroughbred stud, Visor. We had lots of colts sired by Visor." I showed her the picture of her father mounted to play polo. Beside Laurie were three other finely mounted men, sticks proudly held aloft on the Cochrane Polo Field. Model T Fords appear in the background. "Why, that is my dad, on Mayflower – thoroughbred mare, of course, Lazy JL brand on the left shoulder. As a polo player he was known as one of the most stylish horsemen in Alberta."

Laurie worked for Pat Burns, who bought the Bar C Ranch from George Creighton's estate. His job was to gather the horses belonging to the estate. Creighton's dream was to own one thousand horses, but he didn't quite make it, as they only gathered nine hundred or so.

Laurie Johnson and his polo pony Mayflower, 1930s.

The Creighton horses of the Bar C Ranch likely formed another line of escaped horses that mixed with the Kerfoots' and Johnsons' thoroughbreds, which in turn mixed with Stoney stock to make up the wild Ghost herds. The Bar C Ranch borders the Butters grazing lease, which has wild-horse habitat on its western boundary.

I asked Donna if any of Visor's progeny could have mingled with the wildies. At first she was silent and gazed out the window to the horse pasture beyond. I rephrased my question, asking her if any of their dude string horses ever got away to run wild. She said, "Of course, the odd one, and we never saw them again." Approaching my original question from a different angle, I asked her if she thought there was some good thoroughbred blood running in the wild horses near the Butters lease. She answered, "Well, there could be. I saw the odd one."

The Johnson family was good friends with the Stoneys on the Rabbit Lake Reserve and likely traded the odd horse. Erik told me the story of how, when Jean Johnson was homesteading, she had to stay in a tent for ninety days to prove she was living there. The camp-out happened in February, during the 1930s.

The Johnsons' Stoney neighbour Jacob Swampi, also camped out, on band land near Rabbit Lake, worried about Jean being by herself and would ride up almost every day to check on her. He wouldn't get off his horse in the company of a woman alone, though, without her husband present. Laurie had a contract to supply meat for a small lumber camp north of the ranch. One day when it was –40°C, some loggers came by saying they were out of beef. With Laurie away, Jean saddled up a horse and rode up the valley to chase down this fat dry heifer for butchering. She broke trail all the way up and cut the heifer out of the herd. Once on the forged trail through deep snow, the cow had nowhere to go but toward the ranch ten miles away. On the way back, as Erik told it, just as Jean was beginning to think she wouldn't make it, she spotted this huge bonfire beside the trail. Swampi had seen her ride out and knew she would be cold, so he'd lit the fire and left. Jean swore the fire saved her life, and the heifer and the horse too.

◈ ◈ ◈

The possible mix of blood in the wild horses of the Ghost River took another turn when draft horses appeared on the scene in the early 1900s. The Eau Claire Lumber Company put in the Tote Road, parallel to the Forestry Trunk Road, Highway 40, on the east side of Waiparous Creek, to haul supplies to their logging camps. In the winter, they would sleigh logs down the road onto the ice on Waiparous Creek, which flowed into the Ghost, then into the Bow and on to Peter Prince's sawmill on Prince's Island, now part of the city of Calgary. The company of course used draft horses in these logging operations, and undoubtedly some of them ran free, never to pull a log again. There were other draft horses being used in the area as well, which also likely ended up joining the wild herds.

Before Richard Butters married Donna Johnson in 1950, he was living in a cabin on the OC (the ranch where Erin Butters and her husband, Darcy Scott, now live, alongside the Ghost River). Richard and Donna built a new place and the cabin was moved up to the Lazy JL Ranch to be used as a bunkhouse. That's it in the background of this photo of the three generations of the Butters clan who run the ranch today. Left to right are Alyssa, Donna, Erik and Erin Butters.

A cow on the Butters grazing allotment at Horse Lake.

Brenda Gladstone, of the Wild and Free initiative with the Stoney Nakoda people, told me an interesting story on a ride we went on to see some of the wild horses. She said she had mentioned to some Stoney elders that her heritage is closely tied to wild horses through her grandfather, Felix Desjardine. They asked her if she was related to the Desjardines who built the corrals of the same name on the Stoney Reserve near Morley. She said yes, Felix had built those. He came to Alberta from Minnesota, bringing purebred Percheron horses with him on the train. Felix told her about how he had driven about three hundred head of horses and cattle across the 22 Avenue Bridge in Calgary and up to the Stoney Reserve at Morley in 1918. Once there, he let the horses and cattle run wild, rounding them up later in the summer to take them back to the Desjardines' homestead ranch. The Desjardine corrals were built for rodeo and other purposes, and the Stoneys still refer to them as the Desjardines corrals.

In 1919 the Spanish flu was rampant. When Laurie Johnson heard that the Eau Claire Lumber Company's foreman had the deadly disease, he waved down a horse and rider travelling up the Tote Road and sent the last bottle of Kentucky Tavern bourbon from his supply to the ailing man. The foreman later swore Laurie's bourbon saved his life. Not sure what that has to do with wild horses, but it says something about the horses and riders that used the Tote Road – horses as tough as they were, that loved the wild!

A wild horse showing the influence of a draft horse ancestor. Taken from a capture camera of the first herd of bachelors I saw in the Ghost Forest.

Whatever their origins, the wild horses of the Ghost Forest would, for the most part, turn the eye of anyone who knows their horses. However, by some geographical fluke, the Ghost wildies have gone unnoticed by most people and have developed ways of living with other wild animals that I could not have imagined until I began studying them. Their story is one of survival against all odds.

Yes, their bloodlines are mixed. Cuero's, Pocaterra's,

Kit's lead mare has a thoroughbred head if I ever saw one.

Draft-horse bloodlines are also evident in a herd I photographed at Horse Lake in the winter of 2009.

El Cid's and others' value does not lie in their Spanish Mustang blood, if they even have it. Their value is in their ability through selection of the fittest to survive independent of humans and as part of a matrix of other wild creatures. Equally significant is the manner in which they represent Alberta's heritage: the oral stories, histories, scientific accounts and family stories that make up Alberta's history. The wild horses of today carry in them their ancestors, which were a big part of the lives of the early settlers of the Ghost Valley, whether they were First Nations or Europeans – no wonder they are survivors.

 AT FIRST GLANCE, FROM A DISTANCE, I THOUGHT THESE WOLVES
WERE BLACK ANGUS CATTLE – A THOUGHT I INSTANTLY DISMISSED.

CHAPTER FOUR

Surviving through Relationships: Wolves and Wild Horses Interact

When I was riding in the Horse Lake area in mid-September of 2006, I emerged from the forest to see what appeared to be a small herd of black Angus cattle on the west end of the lake. But there are no black cattle on the Aura Cache Permit, just reddish-brown ones. Instead, maybe they were black bears. But bears do not congregate around a lake that only contains minnows, and Rocky Mountain bears haven't learned to fish, anyway. My heart pounding with excitement, I realized these black shapes were wolves. Four smallish black ones suddenly pointed their noses to the sky and howled, and two larger grey ones joined them. Long accustomed now to taking behavioural cues from my horse in such situations, I relaxed and took some photos as Hope quietly gazed at the gathering.

I owned a part-husky dog in the 1980s that loved to howl with the coyotes. As a party trick, I would imitate her howls to get her going. I perfected the smooth shifts in pitch and volume by tipping my head back, circling my mouth and launching forth; at least, my dog and the coyotes thought so. As I watched the wolves, I tried out my howl from Hope's back. All four pups started to

howl again, their song echoing across the Keystone and Wildcat Hills. Of course, I had no idea what I was saying to them, but it was fun. A couple of the pups rose and trotted in our direction. I couldn't see their eyes from that distance, but I hoped they were somewhat astonished when they realized that a human was attempting to speak their language. Looking beyond the lake to the east, I spotted a gigantic black wolf, ears forward, tail up, trotting purposefully toward me. The party was over. I slowly turned Hope around and we walked back the way we had come.

The following week, Hope and I encountered the same pack again, asleep this time, alongside the far shore of Horse Lake. I had clearly stumbled upon a rendezvous site. Before they hunt, adult wolves leave their young at the same selected sites every year, as long as they feel it is safe. Often a sentry or two is left to guard over them. I think the two grey wolves I had seen on my previous visit were acting in this capacity. This time, the pups appeared to be alone. Remaining on Hope, I howled. The pups awakened, raised their muzzles high and answered. Simultaneously I heard a howl of much greater volume

Wild-horse mare and foal skulls.

Wild-horse manure and wolf scat.

from behind me amongst some spruce lining my return route. The hair on my neck prickled. Imagining the photo I was about to bag, I rode in the direction of the howl immediately. Then I saw what seemed to be the same big black wolf, watching me intently from the middle of the wetland. Were there more adults nearby? Surrounding me, perhaps?

Funny what rolls through one's mind at moments like this. I wondered about Charles M. Russell's 1890 painting *Cowboy Sport-Roping a Wolf*. Had the wolf in Russell's painting attacked a horse, or was the roping for sport, or had the wolf killed some cattle, as some do? I have read horror stories of how roped wolves are occasionally killed.

As I remembered Russell's painting, I also recalled something the trapper told me about wolves killing foals. During my six-year study I found a lot of wolf scat, and most of the time it contained deer hair. Occasionally, the scat contained horse hair – usually that of a bay colt. I found an adult horse carcass only once, fully consumed with bones crushed, a sign of a wolf feed.

Even as I thought of all these things, I also thought, "What the hell," and rode across a short stretch of muskeg to a promontory in the swamp, dismounted, took my big-format camera and lens out of the bag behind my saddle and settled to eat lunch and see what would unfold. Rallying my courage, I hoped one of the adult wolves would approach close enough for me to get a good photo.

The big black one howled again, with such volume I almost executed a flying leap into the saddle, untying the halter shank on my ascent. I read somewhere that a hunting wolf howls to bring the young in for a feed. Hope was by then dancing around the tree at the end of the rope. Was she picking up on my nervousness, or did she smell wolf on the hunt? She clearly did not trust my decision to tie her up at that moment. I trust Hope 100 per cent. We left the area. She didn't think much of my request for her to walk slowly through the darkest section of the muskeg.

I later asked Erik for his thoughts on the wolves. He told me how he had ridden into my study area on horseback with Dr. Carolyn Callaghan, who at the time was head of the Central Rockies Wolf Project. Carolyn wanted to see the calf Erik had reported killed by wolves near the sand hills, which are near the west end of Horse Lake above Aura Creek. These hills were the den sites of wolves. When they got there, Carolyn said the sand hills had not been used for dens in years, but that she was sure a den had to be nearby. At that point, they heard a yip. Carolyn stood up in her stirrups and howled. Wolves answered from three different directions and she carried on her wolf-chat imitation. However, they did not see a wolf.

When I hiked into the sand hills in the muddy spring of 2006, the number of tracks made the trail look like a wolf highway, but I never saw a wolf. I felt sick when I saw their former den sites at the sand hills, though. I could

Wolf tracks on the way to the former sand hill den sites.

understand why the dens had not been used. Illegal ATV travel in the area was dramatically evident – the slopes of the hills were torn up beyond recognition and beer cans filled the wolf-den sites. The trail system designed for ATV use and opened in 2006 has been an environmental disaster – many ATV riders do not stay on the trails. Delicate marshland has been shredded, springs turned into muck holes and so forth. And nobody in government seems to care.

Of course, not all ATVers are bad. I've met some who are just fun-loving, decent folk, as confused as I am by the lack of responsibility our government takes for law enforcement in the ATV zone and its periphery in the Ghost Forest. These ATV users are sick of being blamed for the wholesale destruction of the watershed. I maintain that

Illegal ATV use in approach to wolf den sites.

Beer cans thrown into wolf dens.

if a conservation area is to be established for wild horses, the ATV organizations will have to be onside and part of the solution.

At that point in 2006, I wasn't yet aware of the relationship the Ghost Forest's wild horses had with the wolves, but I did know that the delicate balance between all living things in the area was being jeopardized.

A curious incident occurred the following January. Penny, who had first introduced me to the area and its wild horses, joined me for a hike to Horse Lake when the temperature hovered at −15°C and the ground was covered with snow. We wore Yaktrax (like tire chains for shoes) strapped to our hiking boots and carried sharp-pointed walking sticks to help us gain purchase on the slippery terrain. When we arrived at the frozen Horse Lake, I gestured with excitement. Three grey wolves trotted across the snow-covered ice heading east, fleeing suddenly, likely having gathered our scent.

My dog, Cub, had come along with us. She had whined to accompany me when my red truck rolled out, and I gave in, even though I knew better than to take her into an area frequented by predators. Still, I made sure Cub was on leash during our excursion.

When the wolves had gone, we enthusiastically walked out onto the ice to examine and photograph the tracks. Well, we thought they were gone. When we arrived on the ice, I felt as if we were being watched, so I glanced up a short incline to the base of where the spruce forest meets the transitional zone above the frozen lake. A solitary, seemingly gigantic black wolf sat there staring at us. I told Penny, "We have company!"

Her confused response of "Where? What?" came out at about the time I said "Oh, shit!" The wolf was belly-crawling toward us, taking cover behind the odd bush as it advanced. The distance between us wasn't far for an athlete like a timber wolf. I didn't know how to behave around this wolf, which seemed to be stalking us. Relying on my bear experience, I decided we shouldn't run away. I tried chatting, which was something I did to diffuse anxiety when around grizzlies. I also turned my shoulder, appearing to be looking away, which bears do when they want to reduce aggressive actions in their peers. But my tactic did not meet with success. This wolf continued wiggle-crawling at an alarming speed in our direction. I next tried waving my arms and shouting. The wolf stopped for a second or two. We unholstered our Counter Assault bear spray and started a very slow walk back across the lake with Cub, who had not seen the wolf – too busy sniffing scat.

When the wolf reached where we had been standing, he sniffed the ground and stopped the stalk. We headed directly out, looking back over our shoulders constantly, pausing an hour later on the top of Salter Ridge for a cup of tea. Cub's hackles rose and she growled, looking back down the hill. The hair on the back of my neck rose too. We had been followed. We packed up and walked briskly back to the truck, almost bolting for the door, heaved Cub inside and waited. No wolf appeared.

In *Jacobson's Organ and the Remarkable Nature of Smell*, Lyall Watson writes about fear triggering an adrenalin rush to ready a person or animal for action. I have experienced a rapid heart rate, dry mouth and hair standing on end, and if someone looked at me at such a time, I would likely have dilated pupils, too. Lyall writes that all of these things help "set an organism up for 'fight or flight,' making it feel alert and look big, ready for almost anything. The fact that it also becomes sweaty is almost an afterthought, but an important one."

Did the wolf follow our scent of fear or my dog, Cub, who to the wolf was an interloper?

Recalling the "big bad wolf" lore of childhood, I felt afraid for the first time to go into the area alone. I wrote Jeff Turner, whom I had met in 1994 when he was filming a white Kermode bear episode for BBC *Nature* with Charlie Russell on Princess Royal Island off the coast of BC. I knew he had worked on a second BBC *Nature* documentary on the relationship between wolves and wood buffalo in Wood Buffalo National Park in northwestern Canada. I asked Jeff whether my fear of the wolves was valid. He wrote back with some good advice and thoughts.

He told me that if I were alone in the woods, even on foot, I would have nothing to fear from wolves. They might check me out, he said, but any experience they might have had with people will have been negative, so they would keep well away. Wolf attacks on humans are so rare they are thought to be non-existent. Jeff suggested that I think about the likelihood of being attacked by a bear when in the woods, then divide that probability by a hundred thousand in order to calculate how likely it would be for me to even encounter wolves up close.

Doing some more reading on the subject of human / wolf interactions, I had to conclude that the wolf Penny and I encountered was actually stalking Cub. There exist many recorded accounts of wolves preying on dogs. Wolves routinely track and kill lone wolves that enter their range, and they feel the same about wayward dogs. The fact that Cub was loose in her turf angered that wolf for sure. I leashed her the moment I saw the wolf, but I was not aware at the time that she was the cause of the wolf stalking us. Knowing wolves were in the area, I was asking for trouble when I took Cub with us on the hike. At the time, I was ignorant of what wolves might do to a dog in their territory. I have not taken Cub into the wolf / wild-horse area since.

After much research, I gave up my fear of the "big bad wolf" and continued hiking in alone, or I invited Rick or Penny along, or rode Hope in. Hope was perhaps my best partner. Although Rick, as a former wildlife officer in Banff Park, was becoming a huge asset in helping me to decipher what was happening in the Ghost Forest, Hope has infinitely better instincts. Hope could smell, hear and see a hundredfold more than both of us put together. I do not think Rick is offended by this. We all know our limits.

After one of our trips in the area in the late winter of 2007, Rick wrote to friends about the complex relationships between the animals living in the Ghost Forest:

Although difficult to see, there are two sandhill cranes to the left of the bay wild stallion. I spotted them walking with the herd but only managed a blurry image after the herd passed the cranes.

It's not an easy life. We found these two skulls within a couple of metres of each other, mother and foal. Yesterday we stalked three bands of horses and watched a wolf come into the scene and pass close by within sight of the horses. Too cool! The leaves will be out soon, the ground will soften, the land will become a marsh again and we won't see the horses except by chance. But yesterday was grand, and we also had three sandhill cranes filling the air with exceptional loud calls. I got distant video footage of three white-tailed deer watching a stallion mount a filly, all in the same frame. Talk about luck!

◈ ◈ ◈

Sandhill cranes are often in the vicinity of the mineral licks used by the wild horses and other animals. I caught this immature crane on my capture camera where wildies had been only minutes before.

The wolf that roams the Rockies is the grey wolf (or timber wolf), *Canis lupus*. The largest wolf ever weighed in the Rockies was in Jasper National Park in 1945, killed by a park warden. It weighed 172 pounds (78 kilograms). The average male in the Ghost Forest is likely about 40 kilograms. Females are about 2.5 to 5 kilograms lighter. I have seen blacks with lighter guard hairs, and pale greys with brownish to black guard hairs. Colours vary, and the wolves look thinner after winter coats are shed, revealing the colour of their thick undercoats. They look skinny from lack of hair in June and fat in January because of the growth of the thick undercoat and rich, water-shedding guard hairs. I have seen tracks as large as the palm of my hand and in such density that they look like cattle tracks around a salt lick.

At first, I was focused only on studying the Ghost Forest's wild horses. I was not particularly interested in the wolves, having spent enough hours of my life with a summit predator species. But I began to feel there was a mystery to solve related to the wolves, and I decided to open my study to them several years after my first encounters of 2006. It took images from the capture cameras I installed in 2011 to bring the wolf / wild-horse interaction to the fore.

By 2008, saving the area was beginning to weigh heavily on my mind. I knew that logging and increased ATV use loomed in the not too distant future. With that in mind, I successfully pitched a story to Pyramid Productions Inc. for a film, *Wild Horses of the Canadian Rockies*. There is a

Wolf on Camp Ridge, wet and lacking its luxurious winter coat.

shot in the resulting film that all kids love of me picking up a wolf scat, my turquoise nail polish flashing against the snow. The film crew was under oath not to divulge the location of the Ghost wildies, as I was fearful for their safety. Additionally, I required that they bring long lenses and stay well away from the horses. My job was to guide the Pyramid crew into the area for two sessions of filming, one in the summer and one the following winter. When the crew was with me for the summer shoot, they patiently waited three days for a horse to walk on from stage left, as it were. They had better luck and were more prepared for wildlife filming during the winter excursion.

Shortly after working with Pyramid, I zeroed in on an educational venture, Wild and Free, headed up by Brenda Gladstone, co-founder and CEO of the Galileo Educational Network at the University of Calgary. Tracey Read, former manager of Community Partnerships at the Calgary Stampede, facilitated a liaison with Stampede School, run by Anita Crowshoe. Children from grades four to six conducted an inquiry-based study focused on the history and lives of Alberta's wild horses, their relationship to the Stoney Nakoda Nation and the diminishing habitat of the horses. The result was the online resource www.galileo.org. Interviews with Stoney elders, film footage of the wild horses, and students' written and artistic results all formed part of the students' online inquiry.

For this project I collected video footage for the website. Having had enough of being a videographer while on the Kamchatka grizzly bear project, I roped Rick in to help collect footage. He had gone to film school in Toronto before becoming a warden and wildlife specialist; I knew I could count on him! My friend Roger Vernon also helped me obtain the footage I needed for the project. Roger is one of Canada's leading cinematographers, whose work in films includes the likes of *Legends of the Fall*, *Unforgiven* and most recently the *Twilight* series. He is busy, and expensive, but he was happy to help us out and charged much less than his Hollywood daily rate (an understatement) because the project was dear to his heart. I asked Roger why he agreed to help me with cameras and footage. His response did not surprise me with its depth and integrity:

> An area I'd not had the opportunity to explore, it had long represented an untouched part of the Eastern Slopes that I had imagined as pristine. Aside from being an admirer of your work, I got involved because I wanted to see first-hand the condition of the animals and state of the landscape they called home. Little of what we do is simple, but I learned that in Alberta, known for aggressively monetizing its resources, there are many who view the wild horses as simply just another resource to be extracted. I continue to be involved because only public awareness will save this area and these animals.

Both film projects brought me to a deeper level of commitment to my subjects, including the wolves, and brought

Roger Vernon is an expert at hiding capture cameras, having worked on many a set for the movie industry. We were intent on capturing some video footage for the project from heat- and motion-activated cameras. I think his favourite result was a herd of wild horses standing in a lightning storm beside Horse Lake.

me steps closer to asking the bigger questions about how the wolves relate to the horses in the Ghost Forest.

◆ ◆ ◆

My art dealer, Rod Green at Masters Gallery in Calgary, and his wife, Lois, have supported my conservation-related art for three decades. Rod was a marine biologist and had a rare phytoplankton named after him before he became an art dealer. He is considered Canada's leading authority on historical Canadian art, as well as a supporter of contemporary Canadian art. Lois is a sculptress. The patina on her bronze tablets reminds me of the rub trees used by the wild horses – both of them are rubbed smooth, with mottled striations of colour showing under their angular surfaces.

With a fall 2010 exhibition of my latest drawings of the wild horses in the offing, I invited Rod and Lois to accompany me on a hike to Horse Lake. On the way in, Rod spotted a herd on Camp Ridge with his binoculars and with high hopes followed me through ankle-deep water crossing the marsh, squelched through muck and trudged through dense timber – but we had no closer sightings of wild horses that day. Sensing their disappointment, I decided to entertain them with one of my wolf howls. We happened to be standing across the wetland from where I suspected a den site might be located. I howled, and out from behind the dwarf birch popped three grey wolf pups, curiously looking in my direction. After a few yips, back they hid. I could see the bushes moving.

Fox at Deer Lake mineral lick, March 2012.

The scene reminded me of when I had coaxed fox kits out of their den near the cabin I was living in when in Kamchatka. Then I copied the antics of a mother fox when calling her kits forward for a feed. When the kits emerged, I was able to get some amazing photos of them up close.

Now, with Rod and Lois near the wolf den, I did my best flapping grouse sound, wounded mouse and regurgitating fox sound, but to no avail. The pups were in hiding and wanted no part of my imitations. However, I had impressed Rod, who called the gallery with his iPhone: "You won't believe what happened … Maureen howled. She almost barfed. It was incredible!"

A few weeks later, in July 2010, Rick videoed some wildies rolling in the mud at Horse Lake, while I picked off some full-volume RAW images of a muddy wild horse walking by me, with the green water of the lake as a backdrop. The horses were not aware of our presence, so we avoided the usual head-on shots of startled horses prior to their departing high-flung tails and muscled rumps.

Hiking back, as we approached the Sulfur Spring watering hole, we spotted three grey dots moving across the marsh. When we looked through our binoculars, we realized they were wolf pups.

We approached carefully, not wanting to frighten them or scare them away from their rendezvous site. We bent as low to the wet marsh as we could and still move forward, hiding behind one spruce after another as we went. They sighted us from a few hundred metres off and loped with effortless puppy strides across the marsh and beyond. They seemed to be the same pups I had seen with Rod and Lois at Horse Lake, but now a bit older and larger.

A few weeks later, Roger and I sought shelter from the rain as we were gathering footage for the Wild and Free project. We stopped for lunch under some trees not far from where Rick and I saw the grey pups. Out of the corner of my eye I spotted a dark shape emerge from under some willows, followed by two more black shapes – wolf pups. The pups were about three or four months old, and they seemed to be from a different pack than the grey ones Rick and I had spotted earlier. The trio yipped and bit each other's legs in play, oblivious of the rain and us. When one spotted us, it exhibited more curiosity than fear. They all stared, but after a few steps forward, they ambled out into the marsh, one happy blob of furry black after another. They likely knew of the elk or horse carcass halfway across the marsh. Or maybe it was the "Don't run" rule of the wild: if you run, you get chased, and then you get eaten. As predators, they knew: you chase them, they run, eat them!

We crawled closer for some footage and photos. The lead pup spotted us immediately and trotted a few steps forward while its littermate quickly made off with a jawbone. Roger collected some video footage of the pups while I photographed, using my long lens. It occurred to both of us at about the same time that a sentry was likely nearby, allowing our passage. I later speculated that the energy of that wolf's stare reached Roger and me at about

Bachelor after rolling in Horse Lake mud.

the same time, at which point we decided on a slow, respectful retreat.

Shortly after that experience, I devoured Ian McAllister's *The Last Wild Wolves: Ghosts of the Great Bear Rainforest* and felt privileged to have been allowed so close to the pups. In the book, Ian writes that it is easy to screw up trust once it is achieved with wolves. A pack Ian knew well ran toward him once with tails up, ears forward, guard hairs on end. It was not behaviour he had previously encountered, and it made him wonder what negative experience they had encountered with humans in his absence. He knew that if his reactions were not the right ones, he would not have a second chance. The pack was too wary of humans. Later in the book he writes of the "fragment of the trust that once existed between wolves and First Peoples of this [BC] coast is rekindled, that [he is] witnessing the potential for humans to find their place in the natural world."

Back in the comfort of my studio, gazing into the fireplace, red wine in hand, Cub at my feet, I recalled the feeling of the wolf's stare that prompted Roger and me to retreat. The stare, known to most humans as an invisible force, is felt almost instantly. We all seem to know when someone is staring at us, and we usually turn in the direction of the stare. I studied the dingo, one of the closest relatives to the wolf, when I was awarded a resident fellowship at University of Canberra in Australia in 1994. While in Australia I visited Sydney's Taronga Zoo to talk to the dingo keepers and walk amongst the dingoes. I was warned

Wolf pup stares at Roger and me out of curiosity more than fear.

they would circle in front of me to stare into my eyes, to assess my intent. I saw the same behaviour in the semi-wild dingoes of Fraser Island, off the Queensland coast.

Wild horses and wolves alike seem to know the power of the stare. There, in front of my fireplace, I wondered if the wild horses copied the wolf's behaviour, or if the behaviour is somehow in their survival tool kit, a method recalled through the fog of domesticity when needed

Wolf pups with a bone from a carcass found in the marsh.

Myself with capture camera capable of making RAW files, which turned out to be a frustrating project that is still ongoing.
Photo by Rick Kunelius.

to ensure survival. Either way, it seems to be a form of assessment. The stallion Cuero's stare, described in chapter 1, may have been employed to try to gain eye contact with Hope and me. By gaining such contact, Cuero could get a feeling for our intent in his territory and decide whether to run like hell with his band or circle back. Most humans find the stare of a wild animal unnerving. "Don't look a bear in the eye!" is a common piece of advice. I always say it is okay to look animals in the eye as long as your intent is kindly. I have a comical photo of a wolf staring at my capture camera. I imagine him wondering if the big eye is friend or foe.

◇ ◇ ◇

In 2011, when Rick and I were on holiday in southern France, Brenda Gladstone of the Galileo Educational Network forwarded me a letter from Eleanor O'Hanlon, who was interested in my study of the wild Ghost horses. Eleanor lives in France but missed meeting me there by days. After several Skype interviews, she wrote about what I was doing, in her latest book, *Eyes of the Wild: Journeys of Transformation with the Animal Powers*, published in the winter of 2012. Since the completion of the book, Eleanor has provided me with many insights that helped explain parts of what I was seeing with the wild horses and the wolves.

I was privileged to be asked to read not only what Eleanor wrote about me but also the entire manuscript, in which she also writes about wolves. I read with great interest about how Dr. Jason Badridze, the leading researcher of wolves in the Caucasus and professor of ecology at Tbilisi University in Georgia, lived near one pack in the Borjomi forests, where killing wolves is commonplace. These wolves had never trusted humans. Over the course of two years, Jason was allowed to accompany the pack on hunt, often travelling thirty or forty kilometres a day with them, sleeping nearby at night. Eleanor describes how that trust started with the alpha male and female finally allowing their presence to be known, emerging from the forest to stare at him, seemingly puzzled. Finally, Jason made eye contact with them, allowing them a direct assessment of his intentions and thereby achieving trust.

Ian McAllister also reached a place of trust with more than one pack of wolves along the north coast of British Columbia. He writes of how for years he rarely saw a wolf, and only for seconds at a time. Slowly, he began to understand their patterns of travel, feeding, hunting, denning and so on. Ian did not think they would tolerate a human being in close proximity, but eventually two of the packs allowed him to come near. His book is a definitive one on what is beautiful and possible.

Jason told Eleanor O'Hanlon: "It is one of the best feelings in the world when a wolf knows you are not an enemy." After reading about the trust Dr. Jason Badridze and Ian McAllister established with wolves, I realized that Charlie Russell and I had achieved a similar relationship of trust with the grizzlies in Kamchatka. The bears were fearful and elusive until four months of our study passed.

Their response to us seemed to shift overnight. Instead of explosively running away as they had done, more than one grizzly in the dwarf pines allowed me to walk within feet of them as they chewed on cones, seemingly having finally accepted my presence as part of tundra life. It was as if a text message had been sent out to all friends on a social media site saying, "This human is okay."

I applied my reading to my experiences. I thought back to the advice Jeff Turner had given me when I worried about wolves stalking me: the chances of encountering wolves are slim. At this point I realized something was up that I wanted to know more about – I had seen wolves on five different occasions three years into my study of the horses. As well, on my rides in to Deer Lake I often felt I was being watched. I always tried to find who was staring at me, and I searched into the forest for a telltale set of pointed wolf ears. I have not spotted the sentry wolf yet but am sure he/she is there, especially in the spring before and after the pups are born. For some reason the Deer Lake and Horse Lake wolves trust me, and I did not know why until 2012, near the end of my study.

In *Of Wolves and Men*, Barry Lopez puts it perfectly when he writes: "An appreciation of wolves, it seems to me, lies in the wider awareness that comes when answers to some questions are for the moment simply suspended."

◈ ◈ ◈

When bow season opens at the end of August, I see the odd hunter dressed in camo in the Ghost Forest. I feel relatively safe, as bow hunters behave differently from those with rifles and high-powered scopes. However, when rifle season opens in the Ghost Forest for deer around September 20, I end my season. I hate wandering around in bright red and have more than once encountered a hunter with too much booze on board. More frequently now, I spot hunters shooting from ATVs. I am not opposed to hunting game and love a steak of venison or moose, but it is illegal to be in that area on an ATV and off the designated trails. Unfortunately, with no law enforcement, the possibility of apprehending those who make a mess is slim. An officer I know apprehended some dirt bike users in an area illegal for motorized vehicles. One rider turned his bike on him and ran over his legs, breaking them in two places. I also heard of a biologist / volunteer custodian who tried to talk to some men on ATVs in an illegal-use zone. They pulled out a couple of rifles and asked if he wanted to make something of it. There aren't many volunteer custodians.

I ritually howl to the wolves when I enter their area in the spring and when I leave at the end of the season, which usually coincides with the opening of hunting season. In 2011 two different groups answered me. The first chorus came loudly from the north. The second came from farther away to the south. Returning in the spring of 2012, I had barely reached Salter Ridge when a pack started to howl. It is nice to think they knew I had returned and welcomed me, but how could they know from so far away? The howling was likely coincidence, but I prefer to believe otherwise. In retrospect, I also like to think that this howling was a harbinger of the realizations that I would soon gather about wild horses and the wolves of the Ghost Forest. I had been with both for years and was only now beginning to understand how they interacted with one another.

◇ ◇ ◇

Another winter season had passed when I once again spotted a stallion I had come to know: Kit. I had first photographed him in 2007 as a young bachelor at four years old with another young stud, then again two years later running the marsh alone searching for a mare of his own, then with a herd in summer of 2010 and finally alone again the following fall. Kit takes his name from two sources: the first, Rick's sister Kate, who accompanied me into the area one day and was particularly taken by a vision of the black stallion; the second, Kit Carson, a wild frontiersman who explored the western US from California up through the Rockies.

I suspected Kit had been captured by a trapper many times and I think he was some kind of Judas to his kind: he would predictably show up at a mineral lick with a group of wild mares or bachelors, and a trapper would be there. Then the trapper would release Kit and wait for him to return with more wild bounty. Kit seemed unaware of his doubtful relationship with the trapper, being more intent on hooking up with company, particularly mares. I asked a man who traps wild horses near my study area

if he knew Kit, and he said not. But I have heard of other wild-horse hunters on the fringe of the area, so perhaps it was one of those who used Kit as bait.

Regardless, I came to adore the sight of Kit, tall with an ebony-black hide and a polka dot between his nostrils on the white blaze. He front-on eyeballed Hope and me on many occasions, cast sidelong glances other times, but never came close. After a few minutes of circling he always galloped far into the black of the forest, his gleaming black hide becoming one with the shadows except when an occasional ray of sun caught him. Sometimes he would stop to look back, and the light would dapple his form, creating a horse in full camouflage. I completed a drawing of him titled *In Full Camo* that is a favourite. I love it because Kit seemed to always want one last hidden glance of me, or so he must have thought. I always spotted him, though.

As my months of studying the herds progressed, I began to recognize Kit's lone hoofprints in the snow. In early February 2012 I repeatedly saw his track with a single wolf's nearby. It was at this point that I began to wonder about wolf and horse proximity.

I am a track hound. I love to identify track, determine how old it is, consider patterns of movement, numbers of animals or the emotional state of who was responsible – all by the type of impression left embedded in mud. When I am out on trail rides with my friends, I find it unsettling to watch them clip-clop their horses right through fresh spoor.

Kit gallops across the marsh alone. I had not seen him since he was a four-year-old, with another bachelor on the trail off Salter Ridge.

Kit, an elusive black stallion, with a small herd near Horse Lake, 2010.

Kit at Horse Lake, alone again, June 2012.

On every trip into the Ghost Forest, I spot wolf tracks somewhere. Sometimes there is evidence of a pack on the move, but occasionally a single print exists over top of fresh track of wild horses. My first impressions of such tracks were that a wolf was on the hunt, likely hoping to flush a yearling. Incredibly, it would turn out that that deduction was incorrect.

I had read of ravens and wolves. Ian McAllister describes a symbiotic relationship between the two, each benefiting from the other. He describes how ravens alert wolves to carcasses and wolves in turn break open skin that is otherwise impervious for a raven's beak, and so on. Ian speaks of his assistant Chris Darimont's dietary study that analyzed more than 3,300 wolf scats collected from the outer BC coast up into the Coast Mountains. The wolves had ingested many types of birds ranging from ducks to herons, but Chris didn't find a single trace of a raven in a wolf scat. According to Eleanor O'Hanlon, who has read his book *Mind of the Raven*, biologist Bernd Heinrich calls ravens "wolf birds," the relationship is so close.

Spotting the large wolf track alongside Kit's was daunting for me because I do not have a depth of experience with wolves from which to draw. An interspecies relationship within wild animal groups is rare, but it happens and I have been lucky enough to witness one.

I was aware that my many hikes and horseback rides into the area were habituating the animals to my presence – something I didn't want to do. After seeing Kit and the wolf's tracks, I installed my first heat- and

The stud Kit. *In Full Camo,* charcoal drawing, 23 × 29 in., Enns, 2010.

motion-activated camera, which began to record images onto digital cards when I wasn't around. With these cameras I have been able to record a coalition of sorts between the wild horses and the wolves. I have also been able to record some astonishing evidence of their ability to co-exist with other species in the area. Through these images, I have come to a much richer understanding of what I originally thought was a straightforward predator / prey relationship. I began to see the wild horses as members of a wild community, living within a dense and often misunderstood network of relationships.

I installed my first heat- and motion-activated camera near a mineral lick in late February 2012, backgrounded by the Devil's Head, which is located across the frozen marshland. The first download from the digital chip blew my circuits. On February 24, 2012, at 7:09 p.m., two wolves mated in front of the camera and tied, as dogs do, for a short period of time. It is night and dark at that time of year in Alberta, but the infrared function exposed the mating. I was a voyeur par excellence, not that I intended to be. In every pack, only one pair mates, and I had it digitally recorded in the wild – now *that* was a shutter click of luck. Usually the alpha male mates the alpha female. At the time I was pretty sure that two packs had survived the fur trapper of the region, who I had heard trapped six wolves during the winter of 2012. Rick and I had met some wolf hunters earlier that February. They remarked to us that every dead wolf was a live deer or elk (for them to shoot). I was thankful the mating pair had

013F   02-24-2012 19:09:20

Mating wolves caught by capture camera.

survived, and I hoped they were from the group I had already named the Deer Lake pack.

At the beginning of March, the camera recorded a gorgeous grey wolf trotting in and out of the capture camera's field of vision. I decided to buy and install two more cameras in two other areas. I was keen to solve the mystery of the interface between the wild-horse herds, the wolves and the other wild animals in the area, and I knew I couldn't do it when I was present. I am inevitably discovered by a wild animal, no matter how well I perform my camouflage and sneak.

What evolved over the months to come was amazing. For the first three months, I excitedly scanned up to five thousand images per week from the three capture cameras' SD cards. The process became work, but well worth it.

◈ ◈ ◈

I felt as if I had won the lottery when I downloaded a sequence of a black wolf inviting the stud, Kit, to play. Wondering if my interpretation was correct, I went back to Ian McAllister's book and read of wolves and ravens engaged in play. Ian quotes from L. David Mech's *The Wolf: Ecology and Behavior of an Endangered Species*: "I have watched wolves and ravens play together for hours, especially with the pups. Ravens delight in dive-bombing the unsuspecting youngsters, grabbing at their ears in mid-flight."

The images on the next page are from the camera positioned at the Deer Lake mineral lick, taken on April 15, 2012, between 1:30 and 1:40 p.m. If you look at the sequence, you can see the wolf in an ears-forward, tail-down crouch, much like a domestic pup inviting another to romp. A horse approaches in a forward-looking position.

The second photo, taken a few seconds later, shows the horse rotating its axis to reveal the long side of his body. He sniffs the ground, as if to say, "I am relaxed but not in a play or flight mood."

Grey wolf at Deer Lake mineral lick, February 2012.

As Monty Roberts learned from the wild horses of Nevada, horses take flight instantly if facing a foe, but first they momentarily lock eyes with their aggressor in a head-on position. I have seen grizzlies locked for a second in a head-on position, deciding whether it will be fight or flight. The subordinate bear rotates, revealing the long axis of the body, leaving the vulnerable stomach / rib cage exposed. In wild predator / prey body language, this action means "I trust you." Bears will sit and chew grass in addition to revealing their long sides, sending a further message of relaxation. Horses will chew and roll their tongues around in their mouths to show they are relaxed. In the images captured, Kit sniffs the ground, telling the wolf that everything is cool – no one is going to get hurt or eaten this day.

In the third image, the wolf follows Kit, but then, as we see in the fourth image, the wolf lies down to contemplate the view.

Kit regularly moves back and forth between the Deer Lake mineral lick and the Horse Lake mineral lick. A couple of different grey wolves frequent Horse Lake, as do a pair of blacks, one of which could have been the one photographed with Kit at Deer Lake. At this writing, I am not sure if two packs are still alive, or if the cameras are capturing images of one pack and a loner. Regardless, both areas were popular with the wolves for the next two and a half months, until around the time the pups would have been born, in mid-May.

◈ ◈ ◈

Wolf invites Kit to play.

Wolf moves closer as Kit shows trust.

Kit walks calmly away from wolf.

Kit walks off as wolf lies down.

Cuero looking at capture camera, May 2012.

In May 2012 Cuero is alive and with his herd. In the photo above he comes up to the camera at Horse Lake to show his face. How did he know I was concerned? He didn't, I'm sure, but I would like to think he knew I cared.

I was excited to know that his herd, which I had only seen fleetingly in the past three years, was alive and intact with the exception of two blond-tailed studs, colts from the chestnut mare, who had left the herd. A few days before Cuero posed with his herd for the camera, I downloaded some shots of the Horse Lake wolf pack, showing four of them wandering across the mud flat toward their den site.

It was in another image of Cuero's herd, taken at 5:51 a.m. on May 25, of his young foal and three others, that I saw something interesting in the upper right corner that caused me to enlarge that portion of the photo.

I was reminded of Michelangelo Antonioni's classic film *Blow-Up*, which I'd seen in the early 1970s. In it the main character, Thomas (David Hemmings), a bored, wealthy fashion photographer, wanders through a park and photographs a couple embracing. Upon developing and blowing up the picture, Thomas realizes he has photographed a murder taking place in the bushes beyond the couple. He further enlarges the image and thinks he detects a man with a gun in the bushes. *Blow-Up* was considered one of the major films of the 1960s and was nominated for several awards, winning the Palme d'Or at Cannes. It certainly impressed me. I couldn't have known I would use the same technique to unravel a mystery of my own, years later.

When I enlarged the section, I saw Cuero, seeming to communicate with a wolf in a head-on eye-contact position. Whatever the result of the negotiations, Cuero moved into closer range beside his herd. The wolf must have been there, out of camera range, before appearing in the upper-right corner, as the previous picture shows no wolf.

Wolf behind horses (upper right), four minutes before Cuero investigates.

Horse Lake pack, May 2012.

Cuero's herd before wolf walks by.

Cuero's herd shows trust with wolf nearby.

Horses, unconcerned, move in the direction of the wolf.

Subsequent images show the wolf there, in the picture, watching the herd. All of this wolf business is going on in background, with mares and foals seeming to feel totally relaxed in the wolf's presence.

In other images, I see the wolf moving with the herd. I noticed its body language, that of relaxation: tail and ears positioned as were those of the wolf captured on camera with Kit. I wondered briefly if Kit's wolf was the same as Cuero's. But I discarded that notion when I saw that other members of the pack were hanging around the horses. In photos taken a few minutes later, the herd had circled and another pack member walked calmly by, about ten metres from the herd.

◇ ◇ ◇

I wondered what the benefit of a relationship could be for the horses and wolves. Possibly wolves are alarm systems, warning of cougars. After all, cougars routinely prey on horses and foals, lying in the deep grass and debris of the forest, ready to spring for the jugulars of their targets. Wolves, lower to the ground, may be able to detect cougar scent and tracks missed by wild horses. Maybe, as horses graze constantly with deer amongst them, they offer wolves opportunities to kill deer without searching for prey.

As I began writing this book, I wondered how I could delineate a family-type relationship between the wolves and the wild horse. I looked for possible explanations for the wolf and horse behaviour in the First Nations and local non-Aboriginal communities.

Deer and bachelor herd in early June light.

I met Anita Crowshoe when I worked on the Wild and Free project at the Calgary Stampede School. I asked her if she could share a story from her Piikani culture that would shed some light on what I was observing in the relationship between the wild horses and wolves.

It is my understanding that there is a process within First Nations worldview practice whereby individuals are acknowledged and gain the right to tell stories from their peoples' oral history. Anita had been granted that right. First, she told me about the origins of her name and her qualifications enabling her to tell a story relating to my question. Her traditional Piikani name is Sistsi, given to her by All Listens Lady, one of three women who were married to legendary Chief Brings-Down-the-Sun. *Sistsi* means "little bird." In Piikani culture, one's traditional name provides the ability to be acknowledged within First Nations worldview practices. In 1982 Anita received her first transfer, which is the equivalent of a degree or certificate of completion in mainstream society. With her partner and brother, the late Jason Crowshoe, Anita carried the Rider Rattle of the Brave Dog Society on the Piikani Nation in southern Alberta for many years. A person who carries the Rider Rattle is one who has agreed and been accepted to uphold integrity of a specific nature, such as law enforcement, for the culture. Anita is the daughter of a former chief of the Piikani Nation, Reg Crowshoe, and his wife, Rose. She was transferred the Napi and Buffalo teepee in 2004 by Piikani traditionalists who have the right to transfer teepees. Ownership is not a concept in First Nations culture.

In her compelling style, Anita started by referring to the horses as Elk Dogs because their size, mannerisms and ability to run free were similar to the animal the Piikani understood. They used dogs prior to the horse and hunted elk. She told me that she believed the dogs were coyote puppies taken young and raised in the camp. The Piikani also took wolf pups from their dens to become part of the extended family unit. In the traditional knowledge of the Brave Dog Society there is a story of the wolf spirit

Anita Crowshoe (Sistsi Crowmoccasin) in her teepee at the Calgary Stampede Indian Village, 2010.

being given to a human of the Piikani tribe. After this happened, the relationship between the wolf and the people was honoured, and they lived together. Horses also became part of the family unit, roaming freely with the wolves around the teepees.

First Nations worldwide believe in equality between animals and people, which stands in sharp contrast to European philosophy, which places man at the pinnacle. The Piikani believe that the wild horse carries collective knowledge of its honoured position beside the wolf adjacent to the Piikani family unit. The same is true of the wolf's collective knowledge of the horse. For centuries, wolf and horse have been separate, but now, as glimpsed in the Ghost Forest, they are perhaps coming back to share the landscape in a relationship that transcends that of predator and prey.

In his book *Dogs That Know When Their Owners Are Coming Home*, Dr. Rupert Sheldrake, who taught biochemistry at Cambridge, describes something called morphic resonance – also called ancestral memory. He attributes the flight of the millions of monarch butterflies back to the Sierra Madre mountains in Mexico from points far north in Canada to collective ancestral memory. On the way to Canada, the first generation dies in the southern US. Five generations later, a group of monarchs reaches its northern destination. Then the pattern is repeated on the way back to Mexico in the fall; the butterflies travel thousands of kilometres, not one having made the journey before. Sheldrake proposes that their migratory pathway involves an ancestral memory that is inherent in morphic fields. "Morphic resonance is the basis of the inherent memory in fields at all levels of complexity. Any given morphic system – say a giraffe embryo – tunes in to previous similar systems, in this case previous developing giraffes. Through this process each individual giraffe draws upon, and in turn contributes to, a collective or pooled memory of its species."

It was four years into my study that I became increasingly curious about the wolves sharing the same habitat as the wild horses. Ever since 2006, I have heard Darcy Scott, co-manager of Butters Ranch, moan about the cows drifting home from the northern reaches of the lease after a month or so. Darcy blamed it on the wolves. I didn't think much of it at the time, but he also mentioned seeing wolf pups sleeping amongst his cows and calves. I wondered how the wolves could be scaring the cattle home if they seemed to be coexisting so peacefully.

I asked Erik to tell me about the last time one of his calves was killed by wolf up in the grazing allotment. He said it had been years, but he did recall one with a chunk bitten out of its hindquarters and one with a chewed-up tail. He brought the calf with the bite out of its hind end down to my place with its mother to recuperate. It joined some of the heifers he grazes on my land – a safe summer holiday at Ghost Studio Ranch. I was confused, because I had heard of wolves preying on cattle all over southern Alberta. Erik suggested that perhaps his ranch wasn't experiencing this predation because the wolves were preying on the wild horses instead. I believed that, too, at first, especially after I had heard the trapper speak of his need to catch some wild colts before the wolves got 'em! "Run 'em when the spring snow has a crust supporting a wolf but not a horse –

Wolf pups, Deer Lake vicinity, July 2010.

Grey wolf trots by the capture camera, early April 2012.

Cow and bull moose with wild-horse herd.

Grey wolf early in the day, March 2012.

Wildies at Horse Lake watch bull moose pass by.

easy pickings!" However, if you read Lu Carbyn's account of the predator / prey relationship between wolves and buffalo, a different answer becomes a possibility.

In *The Buffalo Wolf*, Lu writes, "Curiously, bison will tolerate wolves in their midst for hours at a time. As if they have not a care in the world, they lie down, chew the cud, and nurse or groom the calves while older youngsters playfully chase one another, running in circles, butting and mounting like adult bulls. At times the scene is surrealistic, like the Garden of Eden. Yet always, the peace treaty is only temporary, and soon Eden becomes Gallipoli."

As Russian ecologist Dr. Jason Badridze described to Eleanor O'Hanlon, "The male ran straight toward a group of gazelles, and the gazelles just stood and watched him! It was astonishing; you've seen for yourself how quickly they flee when they feel threatened in any way. Those gazelles stood calmly and watched the wolf approach because they knew immediately that he wasn't hunting them. They knew because predator and prey are always reading each other's emotions and intentions. They continually study each other. They watch and learn. You see – animals are the best ethnologists!"

Nature movies are full of scenes wherein predators and prey intermingle freely and all is tranquil. Later, when the "need to feed" arises, the scene changes drastically.

As I wrote this book, I realized the verdict is clearly out about the behaviour exhibited between the wolves and the wild horses. I have become relentless with my questioning of those I respect, so that I might come to a tentative conclusion. As I thought about the situation, I was reminded of how I felt back in the early 1990s when I first came to suspect the possibility of grizzly bears and humans sharing a landscape in harmony and even friendship.

Erik described for me a conversation with a wolf expert who told him about good packs and bad packs around cattle. "Somewhere in BC several cow herds graze a long valley, co-inhabiting the area with two wolf packs. One pack kills cattle and the other does not. One pack shows up and the cows don't even get up from chewing their cud. The other pack shows up and the cows are milling about, looking for calves, quite anxious. Cows are dumb, but they know the difference. Something in the cadence of the stride, body language. Wouldn't surprise me if that were going on with the wolf pack at Horse Lake, and I agree with you that it is likely cougars are taking more colts than wolves are."

I don't think the Horse Lake pack is killing horses or Erik's cattle. I have seen too many images of different members of the pack moving through the wild horses, calmness exhibited on both sides. I also do not think the Deer Lake pack kills cows or horses, likely because they are related to the Horse Lake pack and part of their extended family group. This is only speculation on my part. I wonder if there might be another pack in the vicinity whose range the cows sometimes walk into that does prey on cows, occasionally causing them to raise tails and head for home. That could account for the wolf kill some riders found about ten years ago, north of the sand hills.

No answer is clear cut, it seems. I found colt hair in wolf

scat about five years ago but not since. The mystery continues to intrigue.

When I corresponded with Jeff Turner, he told me that when it comes to cattle, wolves wouldn't necessarily know them as prey and how to hunt them. He thought in most instances wolves wouldn't attack cattle, but that doesn't mean they couldn't learn to do so. Jeff wrote that it would depend on what the natural prey base was like in the area. If the wolves had lots to eat, then they probably wouldn't have any reason to look at something strange like a cow to eat. But if they were hungry, that would change everything.

I believe the same can be said of wolves preying on wild-horse colts. Wolves have a natural prey they are adept at bringing down – the deer. There is no shortage of deer in the area, as evidenced on my capture cameras at the mineral licks.

⬖ ⬖ ⬖

In 2010 Lois and I rode to the edge of the forest overlooking Horse Lake. Three herds were gathered fairly closely together. The colts and mares were more or less in the middle, with the three stallions on the periphery, positioned like spokes on a wheel and focusing on something passing through the dense willow not far from where we tied up our horses. Hope and Amigo started thrashing about, so we left the area. I do not know what predator was passing, but there was no doubt in the domestic or the wild horse mind that horse was on its menu. Maybe it was a pack that preyed on horses, or it could have been a cougar on the prowl – but that was unlikely, as cougars are crepuscular animals, usually active at night.

When I spoke to Alyssa Butters about the possibility of some wolves befriending wild horses, I anticipated a look of dismay, but instead she said, "Absolutely." She told me that some coyotes frighten their cattle, while others are respected and allowed to walk close to the cows to catch mice stirred out of their nests by a cow's hoof. She also reminded me of the local rancher of bucking horses, who lost several colts to cougars. They later saw the cougars sleeping off their meal amongst the cows.

The wounds on injured wild horses I have seen and have photos of seem to be of the kind inflicted by cougars, or they may be from frantic gallops while escaping a log trap. I believe cougars are responsible for the cull of an occasional adult wild horse and likely several colts. The wild-horse herds I study are not increasing in numbers, so something is taking them, and it isn't the trapper. His traps have fallen apart in the farthest reaches of the area where I conduct my study; I think he prefers his traps close to a road, for accessibility.

In the end, I circle back to my original question: why are the wolves befriending the wild horses? And I come back again to deer. Deer use the same patterns of grazing and the same mineral licks. Deer are much easier prey for a wolf compared to a wild stallion or a mare set on protecting her foal. Moose are also a prey species of wolves. They graze alongside the horses as well. Moose and deer must have something to do with it.

Cougar on Penny's dead domestic horse across the road from my land.

Wounded bachelor stallion.

Scratched-up stallion.

TWO BACHELORS IN THE MIST BEFORE BEING CHASED
OUT OF THE TREES AND ACROSS THE MARSH.

CHAPTER FIVE

The Importance of Being Wild: Adapting Behaviour for Survival

Deer have provided more to the wild horses than a possible foundation for a beneficial relationship with wolves. I think they have also been "wild mentors" to the wildies in terms of survival behaviour. The first time I saw wild horses in the Ghost Forest, they hid from me as deer would: behind trees, immobile. As I would learn in subsequent years of study, wild-horse behaviour is less like that of domestic horses and more like that of wild counterparts such as wolves, deer and bears. And for good reason.

In order to better observe the behaviour of the Ghost horses, I chose to dress in the colours of the forest; in other words, I took to camouflage. It took a few visits to the Bass Pro Shop, a natural history museum–style outlet for hunters and fishermen, to outfit myself: hat, shirts and so on. But I added to my getup a technique learned from the horses of the bush. When out in the study area, I steady my binoculars behind a spruce tree to study a herd and their behaviour within it. I also test them at their own game of hiding behind trees, using the technique the wild horses have copied from the deer to break up my shape.

I taught Rod of Masters Gallery and his sculptress

Me on Salter Ridge, 2012.
Photo by Rick Kunelius.

wife, Lois, the hide-behind-a-tree trick. On one of our first forays into the area together in 2010, I wanted Rod and Lois to see some true wild horses. It was hilarious watching them practise peeking out from behind a tree, emulating the hiding behaviour of the wildies.

Rod and Lois hiding behind trees as the wildies do.

Rod with bachelor herd stopping beyond him.

Bachelors running through the marsh, chased by something I could not see. They passed within a metre of me in their rapid flight.

We spotted a group of bachelor stallions through the mist on the hike in, but seeing us, they vanished as if into the clouds.

Rod and Lois perfected the hiding technique – when a bunch of bachelors, chased by an unidentified predator, galloped close to our trees, I was almost splashed by beads of their sweat. That was too close, but it was exciting! We realized the stallions were not aware of our presence at that moment, being more intent on what was behind them. When they realized we were there, they stopped not far ahead to stare at us. I suspect the young studs recognized there might be some protection in our vicinity, so they hung around. The predator seemed to have cleared off not long thereafter, as the horses all relaxed their bodies and eventually trotted off.

I mentioned to my cinematographer friend, Roger Vernon, the way in which wild horses have picked up on the deer trick of hiding behind trees. One of his experiences in the field was as wildlife cinematographer for *Running Free*, a film about the wild horses of Namibia. When we spoke, he told me the fictionalized story of a horse, Lucky, who was set free on the Namib Desert. Apparently the horse dug for water and located forage in a manner similar to that of a desert native, the oryx.

Monty Roberts, in his book *The Man Who Listens to Horses*, describes how he perfected his language of equus by experimenting with his use of body language learned from wild horses and the deer on his farm. If humans can learn from the deer their language, it makes perfect sense to me that a horse could learn a deer's body language. My blue heeler pup, Spook, copies the behaviour of my older dog, Cub. If Cub sits before the command to come, Spook does the same.

I often ride with Hope into the Ghost Forest at the first light of day – deer immobile in the shadows, staring at our passage, only evident to me when Hope opens her ears in their direction. Rick and I once watched a mule deer immobile for five minutes – we moved first. We felt the energy from its intense gaze before we saw the animal.

For both horses and deer, immobility seems to equal

Bachelors stop beyond Lois, Rod and me.

Deer, immobile, watching us pass.

invisibility. Horses don't see something if it doesn't move, and neither do deer, wolves or bears. Whether the wild horses of the Ghost copied the deer for camouflage or it was in their DNA to do so is not is not clear to me. What is clear is they have perfected the behaviour over many generations after going back to wild.

Deer, immobile, watching me.

In 2009, while setting up a camera for moving footage near a moose wallow, I spotted a deer upwind from me, out at the edge of my peripheral vision. It bobbed its head up and down. I had seen the wild horses bobbing their heads around as well, and thought at first they were doing it because of flies; I knew for sure it had nothing to do with agreement or greeting! I finally realized both species were stirring up the air. Being upwind, the confused white-tailed deer could neither pick up my scent nor identify this strange creature at its favourite mineral lick. Realizing the circulated air was not helping with identification, the deer circled until downwind, thereby catching my identifying scent. It did not like my smell at all, that human smell that spelled danger. The deer, flag raised and wagging, fled.

◇ ◇ ◇

Having lived around domestic horses all my life, I continued for a long to time expect the Ghost Forest horses to behave like domesticated animals. Somehow when I think "horse" I associate it with all of the behaviours I recognize in the domestic horses I have known. It took a mental leap of faith for me to connect "wild" (and everything that goes with that fraught word) with "horse," even though I had described them as such for years. Time after time I saw how the horses' behaviour compared to that of deer when they encountered danger, showing me a piece of what "wild" meant in terms of the wildies' behaviour.

Kit – Motionless as a Deer in the Forest, charcoal drawing, 23 × 29 in., Enns, 2010.

I have searched for resources on the behaviour of modern wild horses and found precious little, which isn't surprising. We just don't know a lot about the lives and survival techniques of wild horses existing from the early twentieth century to today. It took me three years to find enough consistency in my observations to be able to write about their behaviour, from copying deer hiding and scent-leaving and -gathering to performing loud snorts, from the way in which stallions organize their herds to the way they discipline them, from seemingly inexplicably "polite" manners to increased nocturnal behaviour during the summer months (and other responses to environmental cues). There were many clues along the way, many of which I discovered on horseback.

Cuero Studies Hope and I, charcoal drawing, 36 × 26 in., Enns, 2007.

◇ ◇ ◇

When reading David Cruise and Alison Griffiths's book *Wild Horse Annie and the Last of the Mustangs,* I was struck by a reference in passing: "Part of a wild horse's canniness, according to professional horse runners, was a sort of *moccasin* telegraph that alerted the herds to danger, wrote cowboy and novelist Will James in a 1930s bestseller." The first summer of my rides to locate the wild horses of the Ghost, Hope's sudden halts along the trail confused me. Her behaviour took me three months to sort out.

Hope is a forward-moving horse, always eager to explore new country. She cocks her ears forward in the direction of a deer or upon spotting anything else wild in the forest,

Riding over Salter Ridge looking into wild-horse haven in the fall.

be it moose or bear. Most of the time, after assessing whatever piqued her interest, she walks right on by.

In late August 2006, when I was riding off Salter Ridge along a wild bridle path that would be the envy of any rider out for a hack, Hope halted her free-swinging walk abruptly up to three times in the space of an hour, for no apparent reason. There was no head rotation looking for something in the big beyond, just a full stop. I wondered if she was sick, or afraid. Her behaviour didn't make sense to me, and I urged her forward, reducing the sudden halts to a minimum. I soon forgave her these transgressions, of course; I was more intent on finding some wild horses. As spring rides gave way to those of late summer, her halting behaviour became more infrequent. I should have picked up on what was going on at that point, but I didn't.

In the foothills of the Rockies, frost arrives almost always on schedule in late September, killing the worst of the insects, such as the dreaded horsefly, and making riding trips a whole lot more pleasant for woman and horse. It was a lovely warm fall day, the smell of drying grass and leaves hung in the air. We were on top of Salter Ridge, about to head down the old seismic line to my truck and trailer parked below.

Suddenly Hope halted as she had done earlier in the season. At this point on the trail, her pace usually quickens as she anticipates the removal of the sweaty saddle pad followed by a rubdown and return to the grass of her home pasture.

Curiosity finally kicked in. Something unexplainable was happening. Maybe it was one of the ghosts after which the forest was named!

I directed Hope to slowly circle in concentrically larger

circles. After about four rounds, which were taking me down a steep part of the ridge into some thick timber and undergrowth, Hope stopped again, but this time with ears pitched downhill. I couldn't see a thing other than trees and the odd bird flitting from branch to bush. Focused in the direction of her pointing ears, we remained in one spot, her motionless and me fidgeting a bit. Then I trusted her and froze with her. About fifty metres away, I spotted one black horse's tail after another flip out and back from behind a spruce tree trunk.

Nothing happened for another five to ten minutes. Hope would not or could not move, and neither did I. Then Cuero, the stallion who had stared us down three months earlier, trotted silently uphill to a place behind some low-growing birch, turned and proudly faced us from about six metres away. A younger stud followed. Cuero let forth with a loud blow, causing Hope to stiffen, her muscles quivering. The two studs held our gaze. After no apparent shift in the dynamic he blew again, this time signalling his herd to run. With a crash of breaking branches and thundering, shaking ground, Cuero and his son followed the herd. A bird flew up from near Hope's foot. Only then did she move, and shook all over.

What was it that caused Hope to stop in the first place, up on the ridge? If an upslope wind had been blowing, loaded with horse smell, she would have turned toward it. If Cuero's herd had been moving and stopped, the shift in sound and vibration in the ground would have been obvious to her, but then she would have been instantly alert, ears pointed in the herd's direction. Her stopping behaviour suggested that something else had happened.

We know that some mammals transmit messages that humans cannot detect. Whales and dolphins use sonar transmission. Other animals too respond to cues we cannot sense. I was reminded of the time I was in central Kenya in the late 1980s, filming a herd of about a hundred elephants. They seemed to be fleeing something I could not see, when on cue they all stopped at once, foot after foot held aloft in mid-stride. They likely felt a signal from their leader through their feet or by some other sort of transmission. I also suspect that the grizzlies in Kamchatka were capable of transmitting information over distances, but I still have no concrete evidence. I wondered if wild horses have somehow recuperated an ancient skill – a form of super-sensory messaging that somehow tuned in my mare. Their message did not come from any physical movement, as they were not moving until we reached their vicinity, after Hope's sudden stop.

It is not that domestic horses don't employ any of the signals or behaviours I have seen in the wildies, but this experience with Hope and Cuero's herd revealed to me that the wild horses' silent language is more refined, honed for survival. As well, wild horses are fully capable of domestic-horse behaviour; whether or not they choose to employ those behaviours is another story.

◇ ◇ ◇

Beauty, my sister's hand-me-down, was my first love in the equine world. She always nickered her greeting when I showed up with a carrot fresh from our ranch's family garden. When we returned from a ride, approaching her horse buddies in the pasture beyond the barn, her frantic whinny seemed to fill the valley. Wild horses use the same vocalizations as domestics but with much less frequency, if at all. When I ride into a wild herd's range there is no whinny of greeting if a herd spots my mare – there is silence. Hope never neighs when she sees the herd. She obeys their command to remain silent. However, most domestic horses unfamiliar with wild-horse country will neigh immediately upon seeing one of them. The wildies' response to this is to run in the opposite direction.

Beauty uttered the full complement of vocalizations typical of her lot, the snort being the one that frightened me the most. A snort always meant that something was amiss. It means the same in wildy territory. The loud blow that a wild stud expels is a universally understood warning of danger, and it never fails to cause tension in Hope, and me to flinch. All of the wild stallions behave in a similar manner when alarmed. They hold their heads high, alert and ready for any eventuality. They blow with a loud expulsion of air that echoes across the marsh. Their herd leaves immediately, heading up into the timber, away from whatever their leader stood his ground to encounter.

I once owned a mare that blew like a stallion when she was surprised by my presence. I occasionally tested her response by crawling through the grass in her direction. When she snorted, I performed the loud blow back, causing her to race in circles around me, bucking and rearing in delight, knowing who was there. (It would be stupid to tease wild studs in such a way. Really, if they are anything like grizzlies, they would hate being the object of a joke.) Hope let out a blow one day when we encountered a bull rut emerging from the timber, trotting toward us with his ground-covering stride. I decided it was time to move, but I did hold her steady a few more minutes, emitting a cow-moose-in-estrus sound to take a few pictures as the moose zeroed in on us.

I am often asked if I feel in danger when a stallion trots toward me to investigate and blows. No. I do not feel nervous and have never had a wild stud pin back his ears and charge me. The blow does not signal confrontation; it is a warning that things could escalate into a defensive fight, the signal for his herd to be alert and on the move.

Grizzly bears also expel a loud blowing sound to caution danger, but theirs comes out as something like a woof. The sound comes from deep in their lungs when air is expelled rapidly after a deep inhalation. It seems to echo as it gathers velocity in its exit. When I think about the grizzly blow, I am reminded of recordings of whale "music." The bear's loud blow is a sound of alarm, not aggression. The bear stands on its hind feet for a better view of whatever caused the alarm. In a similar way, wild stallions throw up their heads, becoming taller, more alert, totally focused. The similarities in behaviour between grizzly bears and wild horses never cease to amaze me.

Hope lets out a loud blow as moose approaches us.

Bears, as well as mule and white-tailed deer, use two techniques to gather scent that I have seen used identically by wild horses. The first technique is easy to spot but easily confused with an aggressive behaviour, while the second is totally missed by the average viewer.

While I was working as a photographer for the human / grizzly coexistence study in Kamchatka with Charlie Russell, a female grizzly crossed the river above Charlie and me and slowly circled around, carefully keeping an eye on us. I had my bear spray poised for what I thought would result in a calculated charge. When she arrived at a point below us, she picked up our scent, which she had not been able to gather when she was upriver. Upon smelling us she bolted across the tundra. The fisher bears below were not aware of any of this, and we resumed our photography. I wondered later if this behaviour was related to the famous "bear stalking human" stories I had heard about so many times from hunters, and guessed it probably was, most of the time. Wild studs have now perfected the same trick, so I wonder if there will be stories of dangerous wild horses stalking humans in the Ghost Forest!

Most of us with an interest in the wild have read Farley Mowat's *Never Cry Wolf*. One of my favourite parts is his description of drinking cup after cup of tea so he could pee around his living area to let the wolves know of his whereabouts. Rub trees, stud piles, scent on bushes are the equivalent olfactory signposts for wild horses, wolves and grizzly bears. When friends come with me to change batteries and SD cards in the capture cameras, I ask them not to pee within a couple of kilometres of the camera sites. I suggest they hold it because I know urine is an attractant for wolves, and I am trying to capture untainted images of horses and wolves together. I have photos of a wolf, a few days after we left the area, coming to where Lois had used a bush.

Of course, domestic horses are interested in scent too. The scent left along any given trail in the forest is of endless fascination to Hope. I have no idea what her olfactory abilities are telling her, but I do notice that some of the dwarf bushes along the trail cause eagerness, and others a snort and caution moving forward. I allow her to sniff at the straying branches and the ground because she usually draws my attention to a fresh wolf or teacup-shaped wild-horse track. Hope's collective olfactory knowledge, compared to that of a wild horse with a lifetime of experience distinguishing variances, is for sure minimal but I envy it nonetheless. I doubt if the visuals I live for are as exciting for her. Spotting a wolf track as large as my palm, deeply pressed into the mud, triggers a photo opportunity and a quickening of my blood.

Grizzly bears mark trees as part of their communication system. In the Khutzeymateen Grizzly Bear Sanctuary along BC's Pacific coast, I was once shown a "rub tree." It seemed a sacred place, with complete, beautiful, bear-sized footpads etched in the lush green moss. Every bear approaching this tree ritualistically stepped in the same

Teacup-shaped wild-horse hoof track in the mud below Salter Ridge.

Wolf and black-bear sign in ATV track, September 3, 2011.

place en route to the dedicated tree. The pitch oozed out of the places peeled by bear claws, where the animals often reached up to 2.5 metres above the ground to scratch their bellies and rake their claws through cambium layers. I watched a Rocky Mountain grizzly rub its belly and then its back after the bark was shredded by its claws, leaving grizzly hair sticking out like soft, short porcupine quills.

I found such a bear rub tree in the Ghost area, outside a wild-horse trapper's salt-trap corral, and thought it a strange location. A stud pile was nearby. Other rub trees in the Ghost Forest are located along trails leading to mineral licks and those too are accompanied by stud piles. It seems as if both horse and bear want their presence identified in the same locations. Stallions ritualistically go to a place where the previous stud has left manure, sniff it, turn and leave their droppings. I do not think the practices of leaving stud piles and marking rub trees are saying, "This is my territory." Rather, I would say that both are opportunities to leave "calling cards," or ursine and equine text messages, to use today's terms.

Grizzly rub tree alongside the trail to a mineral lick.

Stallion comes over to stud pile left where ATVS have illegally crossed the marshland. He sniffs, turns and leaves his manure, thereby marking his visit.

Wild horses too have rub trees, used not only to scratch an itchy back in fly season but also to leave scent for other herds. They are easily spotted in the forest if you know what you are looking for. Like a bear's trail to a rub spot, the wild horses' trail is worn and the ground scuffed about beneath the tree. One I have seen is a tree bare of branches and left at an angle after having been blown over by the wind. The tree's surface is polished like ebony, with horsehair caught in the odd rough spot. Riding by this tree, I cannot resist running my hand along the silky smoothness where the tree has been finely sanded by hundreds of horses' backs.

The smoothness created by equine backs over decades of rubbing reminds me of the rub rocks chosen by grizzlies in Kamchatka. There are no large trees on the tundra to stand and claw, or rub, and so these northern grizzlies have chosen large rocks located on promontories for furred rib-cage rubbing. Softer parts of the rocks are indented and polished from the rubbing actions of thousands of bears, smooth as a piece of soapstone in the hands of an Inuit sculptor.

My friend Lois spends hours sculpting clay to capture a wild horse disappearing into the forest. When her tablets are cast in bronze, their surfaces reflect the smoothness created by her fingers as she moulds her forms in a way reminiscent of the way wild horses rub trees. Some of her final patinas are the colours of those rich red-brown-to-black rubbed trees.

I often wish I could gather scent the way my mare Hope does, which is a fraction of the ability honed for survival of a wild horse or a wolf. Scent, whether left on a branch along the trail by an animal brushing against it or by a squirt of urine from a wolf, is an intentionally left calling card. I have seen bears move to the side of the trail to brush their fur against a bush, thereby marking their passage. Hope lowers her head constantly to smell foliage along the trail. Her scent is left, as well. She can backtrack out of anywhere at any time of day over terrain where not a visible sign of her passage exists. She smells her former passage. Wild animals' sense of smell is a hundredfold greater than that of humans. We rely on GPS if lost. A wolf or bear or wild horse survives using heightened olfactory abilities, matched by its ability to hear. I suspect sight is number three on the list of necessities for wild survival but nonetheless also important.

❖ ❖ ❖

Another set of behaviours wild horses have that differs from their domestic cousins out of necessity relates to herd behaviour and organization. The first of these is the lineup. When I began studying the Ghost horses, I soon realized there was more to the lineup of a departing herd than I had originally thought possible. There is more to it than dominant positioning and social order. People who have studied wild horses always say the stallion is last in the lineup. However, as was the case when I was studying the orphan cubs in Kamchatka, I was in for a few surprises when it came to herd organization.

Star-faced stallion stands assessing situation as lead mare with foal heads off.

When raising the three grizzly cubs in Kamchatka, I finally figured out that they chose their travelling order for a reason. Once I'd figured it out, as described in chapter 1, I was amazed by how protected I felt within their lineup rather than at the end. So I probably should not have been astonished to see the lead mare of a small herd suddenly

Stallion in rearguard.

Cuero leading his herd, with mare in foal, 2006.

move to the middle of a lineup with a young stallion behind her and the head stud in front. She was heavily in foal and also sought the envelope of protection that a carefully considered lineup can provide.

The position of the stallion, whether at the beginning or end of a departing procession of wildies, tells a lot about the level of anxiety within a herd.

A stallion usually determines that his role as defender is best fulfilled at the back of the line of his departing herd. The herd knows something is amiss when he moves to that position, and they do not always wait for his warning blow to depart. The lead mare is usually out front. If there are foals in the herd, the lead mare is faster on the uptake and will signal the herd to follow her immediately after she has sensed something is wrong.

One day, Cuero and his herd were movie-set aligned, grazing and resting at the far curve of Horse Lake, unaware a camera was focused on them. He must have decided it was time to move to better pastures or go upland to avoid the bugs awakening in the building heat of the day. He calmly walked off with his mares, a filly and a chestnut with foal following. This was a calm departure.

The herd watches the lead mare and stallion for leadership, and the lineup for departure is organized according to the situation at hand.

❖ ❖ ❖

The adjectives "polite and considerate" in connection with a herd of wild stallions did not enter my mind when I started riding into the wildies' marshland home. I think of politeness as possessing regard for others, acting in a courteous manner.

On one of my first rides into the area, a herd of bay stallions galloped down the trail in my direction, only to veer off sharply about fifty metres from us. They circled, stopped briefly to look and then galloped into the high country of the forest. They did not signal aggression, nor did they rudely approach my mare. I asked a wild-horse trapper from way back if I was in danger on a mare with the wild studs about. He gave me a strange look, causing me to rephrase my question. The answer was that I was safer on Hope with a wild bachelor herd than I was riding through any domestic herd, and that I shouldn't even consider riding through a pasture with domestic stallions. But why? Domestic horses, it seems, no matter what gender, have no herd training, as they do not need it for survival.

Later, I would discover how wise the trapper's words were!

Lois and I were riding through foot-high late-summer grass on my neighbour's quarter section, across the Ghost River from my place. It was a tranquil day, 24°C with light wind rustling the aspen leaves. Suddenly, a blur of horseflesh headed in our direction. I was shocked to realize there were horses on that piece of land; my neighbour preferred it left as a pastoral landscape. Nevertheless, three geldings, a mare and a yearling, kicking and squealing, circled within feet of our horses, which were now ready to bite and kick back. Not a safe situation for horse or rider. I snatched a long branch of willow and spurred my mare directly at the bunch, successfully running them off and through a gate, which I quickly closed behind them. I have experienced similar behaviour when riding groups of domestic horses. Up in the forestry, I have never had a wild horse approach me with such unruly, in-your-face behaviour. They are taught otherwise. Herd discipline is important in a wild environment.

The first identifiable disciplining behaviour I saw that is unique to the wildy occurred in the early spring of 2007. Penny and I had hiked to Horse Lake and were returning across the frozen marsh by Deer Lake. Two bachelor studs were seeking graze, pawing through the frozen crust of snow, the sound of which masked the noise of our approach. When we were about fifty metres from them, the pair wheeled, long tails and manes flung in a whirl of powdery snow. Jets of steam blew from their nostrils in the sub-zero air. I crouched with my long lens focused on the pair. They skidded to a stop totally focused on us. The younger of the two relaxed his tense pose gazing to the right for a fraction of a minute. The older stud, in a blur of motion, reached his head over and bit his buddy on the neck with a clash of teeth we could hear from across the marsh. That got the attention of the horse with the wandering gaze. He instantly refocused on the potential predator (us). Penny and I both excitedly uttered, "Did you see that?" There was no blood on the youngster's neck, but there was hair missing where teeth had raked his hide!

Young Bachelors on Alert, charcoal drawing, 25 × 21 in., Enns, 2009.

Not long after the episode Lois and I experienced with the unruly domestic herd, we were in my 4 × 4 with a trailer full of our horses, their noses poking out the window. I spotted a wild herd at the side of the road ahead. When I slowed the truck to take some identifying photos,

Owl Creek stallion who disciplined his son, and whom I photographed through my truck window as they approached.

a dark bay stallion moved a few steps in our direction, followed by a younger stallion, likely his son. It all happened so fast that Lois and I found it necessary to discuss the incident at length as we rode into Deer Lake later.

The younger stallion made a movement to pass the senior stallion for a closer look at the horse trailer with our mounts. The head stud reared, and with ears flat against his head in anger, pounded the younger stallion with both forefeet, hooves making a loud whack as they met the flesh just below his head. Instantly cowed, the younger horse wheeled and returned to the herd. Clearly this was a violent lesson in not approaching a vehicle, and perhaps also a lesson in not approaching strange horses, even if they are in a trailer.

I feel this kind of discipline explains in part why the horses appear so polite. The wild studs train their offspring to hold back, exhibit what seems to be politeness but is more likely an ability to assess situations, especially relative to fillies. Often fatal injuries occur in a stallion fight, and the wild studs know one is best avoided and that there are better ways to gain a filly than rushing forward.

◇ ◇ ◇

Of course, herd life in the wild isn't all tough love. You may feel that the manner in which I describe "feelings" relative to wild-horse family units is anthropomorphic. Dismiss me if you will, but do not discount what I am about to describe: behaviours which demonstrate that wild horses are caring animals within their herds and even amongst extended family units – inter-herd relations, if you like. Wild horses have a need for a family unit or herd, more so than their domestic cousins do. The herd is a unit designed for survival. Neither domestic nor wild horses were designed to live alone.

When a stallion clobbers his son for moving ahead of him in the face of potential danger, his motive is to protect his son and family. This behaviour falls under the same kind of devotion to family I witnessed when a stud stood protectively over an orphan foal in the midsummer of 2009. At the time, I was with Penny; we had ridden in to Horse Lake and saw a gathering of bachelor stallions mid-marsh. Penny stayed with the horses while I crept forward with my long lens.

I crawled my way through the dwarf birch and willow to a stunning sneak view of the stallions. I knew they would flee if they spotted a human on foot. (I could not ride Hope in because, as you know, she sinks into the marsh.) Not daring to look up and show myself, I made my approach slowly, careful not to place my knee on a twig that could snap and give my presence away. It was a day devoid of wind, the only sounds the hum of thousands of insects and the distant crunch of horses chewing grass. As their munching sound became louder, I knew I was close. At this point, watching me sneak, Penny was aware of where I was relative to the bachelor herd and wanted to call out but did not (for obvious reasons).

Now that I was close to the bachelor herd, I found a line of sight for some pictures and was astonished at what I saw beyond their backs. A stallion was running out of the forest with a mare and a foal, clearly not hers. The mare was heavily in foal herself. The youngster was only a few days old, and as they approached it lay down, causing the pregnant mare and her stallion to slide to a stop. Seeing them, the bachelor herd moved in the direction of the mare and what must have been an orphan foal. The mare watched their approach, and the stud stood almost over top of the foal, not daring to move a foot for fear of stepping on it, while observing the advancing bachelors.

Stud Guarding His Foal, mixed media painting, 40 × 30 in., Enns, 2009.

It was clear to me he would not allow harm to come to his offspring, but he also seemed to know the bachelors. Likely he had been with their herd a year or two previously. The bachelors exhibited curiosity and edged closer but soon wisely halted their advance, lowering their heads to graze on the protein-rich sedge underfoot. They eventually wandered off at what seemed an intentionally leisurely pace.

When the foal's rest was over, the threesome headed up into the timber. I did not see them again until the following spring. The stud was with two yearlings, and his mare was in foal once more. I named the stud Crowmoccasin and his mare Bird Lady because of a story Anita Crowshoe had told me in which she related a bit of Piikani oral history.

In the Piikani culture, families are honoured by brave actions, and this stallion guarding his foal made me think of that. Anita had told me about how, back in the 1700s, a revered tribesman of the Piikani Nation near Waterton, Alberta, heard about other tribes to the south having Elk Dogs. He wanted to obtain some of these creatures to advance his tribe in relation to others. He set out, seeking a chance to raid the Crow Nation for some horses. It was a trip far south on foot and took more than one season. When he finally reached the Crow camp, he took not only horses but a moccasin right from a Crow chief while the chief was sleeping. Crowmoccasin later returned to the Piikani Nation with the stolen Crow horses and raised the status of his tribe as a result. I thought the stallion's valiant action in the face of the bachelors had earned him the name of this revered Piikani ancestor.

As for the stallion's mare, Anita's traditional Piikani name, as mentioned earlier, is Sistsi, meaning "little bird," so I named the mare Bird Lady, not wanting to disrespect Anita by using her traditional name without formal permission.

During the summer of 2011 I was witness to yet another example of caring behaviour amongst wild horses. That summer, Cori Brewster accompanied Rick and me into the Ghost Forest. Rick is a long-time friend of the late Bud Brewster's family in Banff, and Cori had caught my attention with the release of her CD *Buffalo Street*, which went on to win awards. I thought she might be interested in writing some songs about the wild Ghost horses; and I'm delighted to say she did.

From Horse Lake we spotted what we thought was one herd on the top of Camp Ridge. It took a few hours to hike to a position above them without displacing them; thankfully, an upslope wind was in our favour. We sat above the herd without giving away our presence, enjoying a late lunch as I took photos and watched the herd not far below. While we watched, a puzzling series of events unfolded.

First, I spotted a mare off to the left with a foal asleep at her feet. When it woke, the foal wandered over below us to a group of horses of mixed ages, with two other foals amongst the group. A two-year-old of the second herd left his herd to nudge the baby back to its mother. It was

As we ate lunch, we watched the herd below with a foal at foot.

then that I realized the two herds were using the same patch of hill, close to one another but not wishing to mix. The stallions of both herds seemed oblivious to the exchange I witnessed, and soon, one after the other, the two herds moved downhill. There they met a third herd, led by Pocaterra and the Diva, also with newborn foals. All three units seemed to be extended families that realized they were vulnerable with foals in their midst, seeking the proximity of other herds they recognized that were also with newborn foals. They seemed to be seeking protection in numbers, because a sole herd only consists of about five or six horses.

To make things even more complex, there were two groups of deer nearby, as well. The deer barked as we descended the ridge, a sound followed by one of the studs' loud blows. We had no idea what was up until we caught up to them in the aspen below. Somehow someone's foal had been separated in the kerfuffle, but not for long, as its mare and stud came racing up to recover the little chestnut. There was much waiting and nickering, coaxing the newborns along. All three herds finally headed downslope into the darkness of the mixed boreal forest. I will never forget the seeming co-operation between the three herds wishing close proximity and at the same time keeping the units separate.

A year after that affecting scene, I witnessed an amazing display of sensitivity. A buckskin horse was grazing on the north side of Horse Lake with a lone stud that looked like Kit. The light-coloured one stood out, causing me

Foal on Camp Ridge wanders toward another herd.

to believe he was not of these wild-horse herds. Having seen the horse alongside Kit in capture camera images, I thought his build was that of a gelding, his coat and manner that of a domestic horse. I suspected someone had turned him loose, thinking running wild a better alternative than the horsemeat plant at Fort Macleod. Little do such do-gooders know that most domestics lack the ability and endurance to run with the wild herds, which possess skills and muscle finely tuned for survival.

This "city slicker," whose colour must have been a glare to the wildies' eyes, unwittingly left Kit's side and

Foal united with the herd.

wandered over to a nearby herd of six with a bay stallion. The week before, I had watched this same stallion chase off a bachelor intent on stealing one of his mares. On that occasion the bay stallion reared and slashed the wannabe thief with his forefeet, chasing him up into the timber, causing a great crashing of branches. Upon returning, the stallion snaked down his head aggressively, weaving it back and forth to move his herd out of the area. This time – to my total surprise, as I was expecting another fight – El Cid nudged him back to his waiting companion, ears forward, not flat back against his head in anger. Perhaps the stud realized that the buckskin was vulnerable, that he was a gelding offering no threat. I saw the buckskin once more a week later but not again, and Kit was running with a filly two months after that.

I suppose one does not often think about stallions as being "caring," but I must say that after seeing the bay stallion with that gelding, and Crowmoccasin with Little Bird and the foal, I have started putting the words "caring" and "stud" together regularly. At least when it comes to wildies. But, as if to impress upon me this protective behaviour, the nature of the stallion to care for his herd, another example of a protective stallion came up as I was writing this book.

Driving north on Owl Creek Road in mid-June 2012, I spotted the Owl Creek herd. Not having seen them since the year before, I was delighted to find that they were alive and had outsmarted the horse trapper. The stud is a proud dark bay with a white star, the signature of a Ghost

El Cid (on the left), before chasing the buckskin back to the bachelors.

horse. His two mares, a yearling and a young foal headed immediately into the timber to the east of the road with him in hot pursuit, so I had no photo opportunity.

Two weeks later, seeing the same herd, I was sorry to see no foal. To my surprise, the stallion stood as if watching my approach. The others of his herd were already heading up into the timber. A big bay mare appeared, lumbering up out of the ditch on my left side. The skin on her jaw was hanging by a chunk of flesh and her head was covered in blood. Her cheek from her eye to the jawbone was

Owl Creek stud with injured mare uphill as he waits to guard her departure.

exposed, the bleeding flesh shining in the early light. The stud patiently waited for her to join him, and then he led her away after the others. I admired the stallion's desire to wait for the mare, but I thought she would probably be dead within a week, vulnerable as she was to predators.

However, another week passed and once again I made my way to change batteries and SD cards in my capture cameras. Approaching the bend to the Direct Energy gas well site, my truck and trailer wakened a mother grizzly and two cubs from their snooze beside the road. She reared to see us better before leading her cubs into the timber on my right. Navigating the corner, I again saw the Owl Creek herd, and was surprised to see the injured mare not only still alive but looking stronger. The small herd was no more than fifty metres from the grizzly family, which clearly offered them no threat.

I saw the same grizzly family the following week, in the proximity of the Owl Creek herd. The bears quickly disappeared from my line of sight. Driving home at the end of the day, I spotted the Owl Creek herd, again near where I had seen the grizzlies earlier in the day. The stud stood watching my truck and trailer as his herd with the injured mare raced up Salter Ridge.

I haven't figured out the relationship of the mother grizzly and her cub to the Owl Creek stallion and his herd with the injured mare. I saw no indication in either situation of a predatorial situation evolving.

Four weeks passed and I did not see the grizzly or the herd after that, but by then the cattle of Butters Ranch were out on the grazing allotment. The two-hundred-plus cows and calves do a good job of reducing the fire hazard with their browse of the thick, drying grass, leaving little to hold a horse herd. A good thing for the herd's survival.

I find the topic of the emotional spectrum of wild horses as fascinating as the feelings I saw expressed by the three grizzly cubs in Kamchatka. Others have found similar fascinating scenes when observing family units of wolves. Eleanor O'Hanlon told me about the families of wolves in the Russian Caucasus she heard about when she spent time with Dr. Jason Badridze: "Their close family relationships are among the most fascinating aspects of wolves' lives for me. These powerful predators, whose bodies and instinctive responses have been honed by evolution to give chase and kill, take hold of flesh with their powerful canine teeth and rip their prey, live in families where there is intense loyalty and support. Group harmony is maintained through physical displays of affection and demonstrations of deference and respect toward older and more influential members. Wolves are very skilled at communicating emotions through body posture and facial expression."

I have always wondered where the caring nature of the wild horses evolved. I see signs of concern exhibited amongst my domestic horses. Lois's gelding, Amigo, was injured this spring of 2012, and my mare wouldn't let him out of her sight during his recovery. Most horse people can cite examples they have seen within their domestic horses of how they care for one another, to a degree. With the wild herds, kinship and family units are critical, as they all need one another as a unit for survival. For example, the more eyes and ears there are, the better off the herd is when a cougar approaches. The more bodies there are that get along, the warmer it is in winter when they huddle together.

❖ ❖ ❖

Over the past six years, from about the end of June onward, the bachelor herd moves into the area between Salter Ridge and Deer Lake. By this time, the herds with mares and foals have gone over to Horse Lake or deeper into the valleys and higher up the slopes of the undulating, treed landscape. No one sees them up there unless one knows where to go. There seems to be only one bachelor herd that loses and gains members annually. El Cid was in that herd as a four-year-old, but he left to wander with a single stallion; the pair was obviously looking for mares.

El Cid was the leader of the bachelors when they first all galloped toward Hope, and he was in the group when they all crowded around one tree to gawk at Hope and me in 2006. His fellow bachelors have also dispersed, with one or two forming herds of their own. I do not know what happened to the rest over the past six years.

One young stallion I have been able to track is a blond-maned one that I first saw as a colt in Cuero's herd, running beside the light-chestnut mare. I later saw him as a yearling with his blond mane and tail more pronounced.

Bachelors sparring.

When I taught Rod and Lois the hiding-behind-the-tree trick, Cuero's young stallion was in the lineup of bachelors that stopped in front of them. It has been fascinating to watch the progression from colt to yearling to bachelor.

During the summer of 2012, the bachelor herd spent time at the Deer Lake mineral lick, where my capture camera was located. Now there were two stallions with blond highlights. The process seems to be one of first joining up with the bachelor herd, staying with it until they assume dominance and then leaving, either in pairs or singly to find a mare. Within the boys' club there is much play-fighting, as is clear in my photographs. At the same time, the bachelors seem to be practising behaviours that will serve them well when they strike out on their own.

There appears to be a high level of loyalty within the bachelor group. As described earlier, they huddle behind a tree as one unit to avoid detection. They follow the lead of the head stallion without exception. They move as one unit. When they meet Hope and me unexpectedly, the head stallion steps forward to check us out, with the others respectfully behind him in some kind of rank and order. At the same time, they are full of curiosity, seem to love life and are full of energy, like teenagers. I love seeing them most when on my horse. (If I am on foot, the stallion group responds differently.) When Hope and I are together, the first member of the bachelors to spot us throws his head up, ears and eyes on us. The rest of the group, following his cue, see me and with a rapid spin on their haunches gallop into the darkness of the forest, dissolving into ghostlike apparitions of their former selves.

◈ ◈ ◈

Often at Horse Lake I hear ravens screeching and see them circling above the trees. Hope becomes anxious, and I have a hard time keeping her to a steady walk as she shakes her head in frustration. I have no interest in her

Ghosts of the Forest, charcoal drawing, 21 × 30 in., Enns, 2006.

running and attracting whatever is out there. I also hate bouncing my expensive camera, which sits in its leather nest behind my saddle. However, clearly something is amiss in the back of beyond – a kill, perhaps, and a predator near the carcass. Additionally, at these times I am sure Hope can smell something. Curious as I am to see what she is sensing, I resist the temptation. I contemplate howling to see if a wolf answers, but I agree with my horse: time to go.

Both wild horses and grizzly bears come to an immediate alert when they hear or feel the sound/vibration of a running animal. The horse exhibits a readiness for flight. The grizzly displays a readiness to chase the rapidly moving object or run with the runner for the fun of it. I was often with the three Kamchatka grizzly cubs when they stood up hearing/feeling the sound/vibration of another bear running. They did not know if there was a predator around causing another bear to flee or if the salmon were running. Making an instant decision based on information I could not interpret, they either fled or followed the sound of paws pounding on a packed mud trail.

It's a real piss-off when I am in the right spot at the right time to take an outstanding photo of a wild horse only to have the squirrel overhead or the bird by the lake let loose with a raucous warning. Rick and I once hiked through some ankle-deep swamp in thick willow for the better part of a day so as not to disturb a herd. Finally, when we circled back downwind, I spotted a squirrel about to chatter overhead. Pocaterra and the Diva were right there,

illuminated by the low-angled winter sunlight. It is the best light of all in which to photograph a sculpture or three-dimensional subject – direct overhead light causes details of form to flatten. Musculature on a horse's body needs light from the side, and on that day the light was perfect. Glaring at the squirrel, I whispered that it keep quiet, and by some fluke it did. Pocaterra was glorious as he moved off in an extended trot.

If you want a wild ride, mount your horse when a thunderstorm is building. Horses seem to respond to the pressure changes, as does my dog Cub, who hyperventilates at the approach of a storm. Hope and I were once caught in a storm that was close to tornado velocity. Golf-ball-sized hailstones were bouncing off the ground. It is not only the pressure change but the pain of hailstones hitting their hides – not to mention the high winds and the noise of thunderclaps – that set animals on edge. A wild horse heads for the deep forest for protection when storms occur. They cannot hear a predator's approach or the other warning sounds such as the shrieking of a bird. They remain immobile, waiting until visibility, and the ability to hear something approaching, return. I had a video camera set up on the edge of Horse Lake that gathered some surprising footage of a group of wild horses during a thunderstorm. They wisely remained in the open at the lakeside, not under a big tree, as lightning flashed across the sky, illuminating the herd and the lake.

Another distinctly wild behaviour of the wild horses is one that I only picked up on after installing heat- and

motion-detecting cameras in the Ghost Forest. It is also a behaviour seen in deer and the other creatures of the area. In February 2012 I was stunned to discover that the horses were nocturnal, and more so at certain times of the year than others. I found that by August the percentage of animals moving at night had doubled from 25 per cent activity in February. Out of 2,020 images taken over three weeks in August, nine images of wolves were exposed at night compared to two during daylight. With the horses, mule deer and moose, only 560 were taken in daylight; about 75 per cent of their activity happened in the dark of night.

As you can see, wild horses seem to pick up many environmental cues, tailoring their actions to what they see, hear and smell. With their ability to read their environment, and their adaptation of the behaviours of their domestic ancestors to ones more suited to this habitat, the wildies of the Ghost Forest have become a natural part of the area. They have gone back to their wild roots and reclaimed the behaviours necessary to wild creatures.

Pocaterra moves off at an extended trot while the Diva watches me take his photo.

KIT ALONE, 2009.

CHAPTER SIX

Not Feral but Wild, Strong and Sacred

✳

With a notably downward, harsh tone of voice, many ranchers and individuals ensconced in the government offices of Environment and Sustainable Resource Development (ESRD) in Alberta describe the Ghost Forest's wild horses as "feral." They might as well be referring to a rat invasion.

In correspondence with me on June 22, 2007, Ted Morton, minister of what was then known as Sustainable Resource Development, wrote:

> The Alberta Government considers the horses that are in the Foothills region to be feral horses that have escaped, and in some instances been released, from domestic herds over the past century. Even though they aren't deemed wild horses, their use is integrated with other uses and values of the landscape.

But what are these "other uses and values"? Minister Morton did not elaborate. I think: logging, recreational use by humans. Given the wholesale acceptance of wild-horse roundups by individuals without licences, I can only assume that the value of these horses, kept wild and in their environment, is of little significance to SRD.

A friend of mine, Christine Wiesenthal, a professor of English literature at the University of Alberta, interviewed Dave Ealey, a spokesperson for SRD. He told her, "They have a legitimate designation now. They're feral horses." Christine wanted to know why they were not considered wildlife. Ealey essentially paraphrased what Christine had already read on the government's "Frequently Asked Questions about Feral Horses" web page: "The horses are descendants of escaped domestic stock, mostly from early twentieth-century mining and logging operations. They are an introduced species, not indigenous to the area, and therefore not truly wild. Moreover, as stray animals, they are pernicious invaders, injuring grasslands by overgrazing, competing for food with other native herbivores and domesticated stock. And they are a road hazard, too…. There are public safety concerns to consider here."

In an essay resulting from that conversation with Dave Ealey, and others, Christine puts a terrifying spin on the word "feral": to have reverted from domestication to a wild state. Christine discovered "a buried, largely forgotten, original meaning. Once upon a time, the word was synonymous with 'fatal.'"

From what I have observed in the behaviour and physical attributes of the horses of the Ghost, "feral – not true wildlife" as description of the creatures falls ridiculously short of the mark and denigrates them. Not only that, a label like "feral" is enough to devalue the horses in the minds of a government and a people, both of which should instead be striving to protect them. In order to bring good sense to the discussion of wild horses, I have studied them for six years, coming to a variety of conclusions about their behaviours, their role in their immediate environments and their value to Alberta's cultural heritage. It is my hope that when the horses are better understood by all, we will be able to protect them and their environment.

◈ ◈ ◈

I have worked out some definitions for the spectrum of horses that I see on and near the Ghost Forest wild-horse turf. I see four categories of horse in the area, which I will describe here to avoid confusion: domestic, free-roaming, feral and wild.

Taking a drive through Butters Ranch, the Rabbit Lake Indian Reserve and on up the Owl Creek Road on Crown land is a good way to observe all four categories of horses, with the domestic horse at one end of the spectrum and the wild horse at the other. In the middle are the free-roaming and the feral. However, please keep in mind that these categories and scenarios I've devised here are – as is the case with all generalizations – simplifications.

The domestic horse is easy to spot. As in seen in my

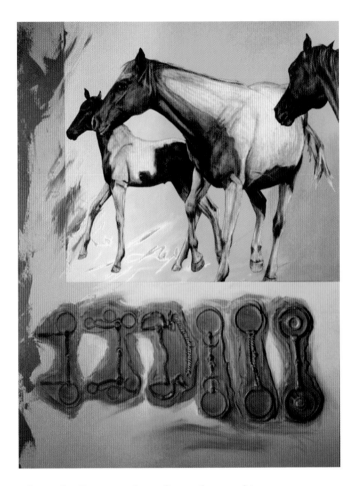

Above the Bit, mixed media and cast objects on canvas, 60 × 48 in., Enns, 2005.

Camargue horses at the breeding station, Tour du Valat, Camargue, southern France.

Generations later, some of the wild horses of the Camargue are bred for tourist rides in southern France.

painting *Above the Bit* from my series *Equestrian Heroes*, domestic horses are bred to be ridden. These horses reside within fenced areas and are shod all around for ranch work. On my way out to the Ghost Forest, I stop my truck and trailer on the side of the road, near some horses in pasture. The fenced-in horses nicker and trot over to investigate my horse, who pokes her nose out from between the slats of the trailer window. Looking beyond the curious horses, I see round bales of hay beyond their paddock, ready to be forked over the fence for their winter feed. These animals are domestic stock.

There is a group of famous domestic horses that have been labelled as wild when they are not: the so-called "wild white horses of the Camargue" in southern France, at the mouth of the Rhône River. They are domestic in every way. Their ancestors likely ran wild, but now, for touristic reasons, they are raised within fenced enclosures, kept close by for pony rides, fed when necessary, blanketed when cold, given vet care when sick, and many are shod. Select groups are kept in large grazing areas and are used for breeding the "true" Camargue type. Visitors stop their cars to pick grass and feed them through the fences. Rick and I visited the Camargue in the spring of 2011. After a lot of driving, we found one group of three geldings wandering outside a fence. I walked up to them to take their picture. They are beautiful animals, but they are far from wild. Just because an animal has a wild "history" does not make it a wild animal.

Leaving the Butters lease land, I rumble across the cattle guard and enter Rabbit Lake Indian Reserve, easily spotting a couple of free-roaming horses belonging to the Stoney Nakoda people. They cross back and forth over rundown or nonexistent fences. Often a stud is amongst them. The adults are usually branded, and colts run beside their dams. Their colours range from chestnuts to bays, blacks, greys, pintos and the mandatory white horse with spiritual value. When they see me, they look over but quickly satisfy their curiosity: mine is just another truck similar to what they see every day. They return to graze the lush grass. Similar bands of free-roaming horses can be seen walking through the schoolyard or beside the highway at the Morley Reserve, another of the Stoney Nakoda Nation's reserves, located on the Bow River west of Cochrane. It is unlikely they are fed hay in winter; instead, they are left to fend for themselves amongst the dwellings of the reserve, much like their forefathers did in the vicinity of teepees prior to the Stoneys being moved onto the reserves. That said, most of them can be caught with a shake of the oat pail.

Upon crossing a second cattle guard out of the Rabbit Lake Reserve, I see a herd of twelve mixed ages and genders led by a big chestnut stud. They are a bit more alert to the approach of my truck and trailer than the free-roaming and domestic horses are. Still, most of them ignore me when I stop my vehicle. If I get out to take a photograph, one or two young fillies will trot off, looking back constantly to see if their departure will be copied by the rest of the herd.

These are likely wild fillies the trapper caught and let go with his herd to cross them with his chestnut stud. Most of the herd continues to wander about, used to people, having only run free in the forest for short periods of time. Clearly one or two in the herd could be caught with a pail of oats as a lure. These are feral horses, some of which I believe are owned by the trapper. They are grazing the Aura Cache Permit, not one having paid his way! The herd is likely rounded up once or twice a year by the trapper, with some taken out to train for the saddle.

One of Butters' cows alongside a feral horse on the ranch's grazing allotment, photographed through truck window.

These feral horses, grazing on the Butters Ranching Ltd. allotment, show none of the signs of alertness of their wilder cousins.

Understandably, these horses piss off the rancher when he turns his cattle up into his Forestry allotment for the summer. By this time, blossoms of clover and seed heads of native grasses have already been nipped off, and some of the lush graze for his cattle has been trampled.

These truly "feral" horses may have escaped from ranches or been turned loose. Alternatively, some of them may be horses that have grown up in essentially wild conditions but have developed a comfort level with humans due to being fed by trappers or well-wishing photographers or tourists. The wild horses can regain domestic tendencies, which is why I have been so careful to keep the herds I study independent by relying increasingly on capture cameras. The length of time a horse has run with a wild herd or the number of generations its forefathers have run wild will have an impact on how fast it will redomesticate.

In June of 2009 Penny and I saw a heavy-boned grey stallion that was clearly not a wild animal. The stud trotted up to within six metres of us before casually walking off. As I understand the story, the trapper turned the stallion out on the Butters grazing allotment – sometimes also wildy territory – so that any wild fillies he had his eye on could be more easily lured into a salt trap, with the stallion unafraid. I spotted the stallion near Horse Lake with a wild filly at his side and later alone near where I park my truck and horse trailer. I never saw him again. I think he was likely sold to one of the horsemeat plants near Fort Macleod. It is legal to process horses for

This newly released stallion was meant to run with wild fillies but did not last, as he was run off the range by the wild studs.

human consumption in Alberta, but all of the "product" is shipped overseas because there is no longer a market for it here. (Formerly it was used for pet food or domestic mink feed.)

Another group of feral horses hang around on the edge of Rabbit Lake Reserve and on into the forestry, along Owl Creek Road. There is a big chestnut branded stallion, several grey two- and three-year-olds and an assortment of mares, some of which look a bit wild. I once asked Anita Wildman if any of these are the Cody

Feral herd, possibly managed by the trapper, wanders from Crown grazing land onto the Rabbit Lake Indian Reserve.

Wildman family horses. With a look of disgust, she said they definitely are not. "They are the trapper's horses. I see his horses near the reserve every year. His horses are consuming free graze on both Forestry grazing lease and reserve land." The trapper in question is a nice guy and a good horseman, as described in the first chapter of this book. I speculate that everyone allows his grazing transgressions in the spirit of the "Old West." At the same time, I hear frustration about these horses from not only Anita but also the Butters.

Penny, riding Rocky, watches a feral proud-cut gelding.

In the summer of 2011, a star-faced bay ran up behind Hope as we were entering the Rabbit Lake Reserve, coming so close I took a branch of willow in hand to beat him off. He exhibited a total lack of respect, as most domestic horses do when encountering a strange horse – kicking, squealing, coming in close. He followed us about two kilometres before whinnying. Another horse answered from up the hill. Watching him run off, I took him to be a gelding. Penny thought he was a stud. Seeing Anita Wildman at the entrance to the reserve, I asked if she

knew the horse and laughed at our different evaluations of his gender. He was theirs, Anita said, and a "proud-cut gelding," which is an odd condition about which there is some debate. Even though these geldings are neutered, they still think they are stallions. I believe the Wildman family caught the semi-stud, as I did not see him on wild-horse land again. I continually recall the differences between his behaviour and that of the wildies – he showed no respect for our horses at all. He reminded me a bit of the palomino I saw running with a wild stud at Horse Lake in 2010. The horses that have escaped or been turned loose behave differently.

On three occasions I have spotted horses that by colour and behaviour stood out within the wild herds. Sometimes people release a favourite horse they cannot bear to put down or can no longer afford to feed. Sometimes the plant in Fort Macleod is full to capacity. The newly free horses usually don't last long in the wild. I watched a stud put the run on a buckskin mare and saw a grey stud get run off by a different stallion. A gorgeous palomino filly ran with a young stud for about two weeks and then disappeared. Were they subject to predation? I have no way of knowing. Black, bay and dark chestnut are the predominant colours of the Ghost wildies – and for good reason. As well, the released horses do not have the skills the wildies have – the relationships with other animals, the tool kit of wild behaviours – to survive in the wild.

The only truly biologically wild horse left in the world

A four- or five-year-old colt that either escaped or was released in 2009.

This newly released or escaped buckskin did not last long with the wild herds.

today is Przewalski's horse, also called the Takhi, a Mongolian breed. According to the International Union for Conservation of Nature (IUCN) Red List, the Takhi was listed as extinct in the wild until 1996, when its designation was changed to critically endangered after reintroduction. The association for the Przewalski's horse, TAKH, was founded in 1990. Its aim is to create a breeding centre for free-roaming Przewalski's horses on a plateau in the Lozère region of France, as well as to reintroduce some of the horses born there to Mongolia, the historical range of the breed, from which it disappeared at the end of the twentieth century. Reintroduction has taken place in Mongolia, China and Kazakhstan, with nearly four hundred horses now roaming in reintroduction sites in Mongolia and China. While this horse is genetically different from domestic breeds – Przewalski's horses have two chromosomes more than domestics – by some definitions the Takhi is considered feral. A feral organism is one that has transitioned from being domesticated to being untamed. I view zoos and breeding facilities as a form of domestication: the animals rely upon and are cared for by humans.

So, are the Takhis released into their historic ranges feral or wild? What are the differences in behaviour between feral and wild horses? I would suggest there are marked differences between the two. Before we finish our drive through the domestic / wild spectrum of horses, allow me to elaborate.

❖ ❖ ❖

During the summer of 2012, with hot, dry spells occurring more often than normal and more bugs than ever, I valued the opportunity to invite Penny to ride in with me to keep an eye on our horses, which were tied up, while I changed batteries and SD cards in the capture cameras. From the images I had gathered so far during the summer, I had noticed that the wolves were not moving about much in the heat of the day, and neither were the wild horses. I picked up the odd night shot of an adult black wolf passing by. I wondered, examining the digital image, if it was the same wolf that in other images had appeared to befriend the wild horses. Summer coats on wolves change their appearance, and I could not tell.

After heavy rain all night, and against our better judgment, Penny and I headed for Owl Creek Road, my truck in bull low to haul my trailer through deep puddles of water along the road past Cody Wildman's place on the reserve. Bumping over the cattle guard, I spotted the first herd of Butters Ranch cattle, their hooves sinking into the mud along the road. It felt like I was ploughing the road with my horse trailer, which was at times up to its axle in mud. Stopping at the top of a hill – and not daring to have done so up until that point, for fear of slipping off the track – I pulled off the muck onto harder, higher ground. A grader for Direct Energy had been by within the last two days, moving a lot of dirt and gravel around, and now muck with foot-deep trenches down the middle of the road marked its handiwork. We saddled up a couple of kilometres short of our usual spot and reached

Salter Ridge as the sun came up, which, thankfully, dried the road enough by mid-afternoon to turn the truck and trailer around and head back.

Looking intently through the aspen as we approached the Deer Lake mineral lick, I spotted a black shape with two pointed black ears. I realized it was a wolf watching our approach to the lick. Then it was gone.

Riding around to the next peninsula of hard ground on Deer Lake marsh, we saw Kit with a bay filly nearby. Penny and I rode within fifty metres of him, which surprised me. Still, through the years, I have discovered Kit to be less afraid of humans than the other wild horses are. I estimated he was about nine years old that day, having first seen him as a bachelor four years before. The filly at his side mothered up to him in her fear of us, rolled her eyes at our approach, all the while keeping Kit between herself and us. Nonchalantly, Kit rotated his body so his long side was to us and ate grass, which in wildy language signals trust.

Then, I wondered, could the wolf he trusted still be there, beyond us, acting as sentry and giving the safe passage signal with its body language to other wolves and thus to Kit? I was so intent on photographing the differences between Kit's behaviour and his filly's response to us that I did not look behind us, across the marshland. In retrospect, Penny and I both recalled our horses intently watching something beyond our field of vision in the trees about five minutes prior to reaching Kit. I suspected it was the wolf and got my camera out but saw nothing. Riding on, I gave it no further thought.

As I mentioned previously, I have always suspected Kit had been used as a Judas horse, caught and released over and over after he came back to a salt trap with new bounty for the trapper. His role could be similar to that of the big grey feral stud in the trapper's herd. Still, I have grown fond of Kit and his haughty appearances with different lady horses, despite my feeling that he is not truly a wild horse like the others in the Ghost Lake area. Following is a series of photos of Kit taken over a four-year period.

The horses that behave like wild animals – that I consider wild, not feral – are the ones I have studied and that over the past six years have shown me behaviour that is very different from that of the ferals. There are very few of these herds left in Alberta. The last time I took account, the area I have been studying has about six wild herds.

The critical part of my definition of "wild" as it relates to horses is the ability to survive as wild animals, in the same manner as the remote ancestors of their species did thousands of years ago. I've already talked about a few of the characteristics displayed by the Ghost Forest wildies that lead me to believe they are truly wild. They live independent of humans. They have adapted behaviours similar to those used by other wild animals: for example, the same method of avoiding detection in the forest as deer. They co-inhabit the wilderness with other wild animals, such as the wolf, having worked out relationships with other creatures that are critical for their survival. They submit to the natural cull of the wild, and only the strongest survive.

Kit with small herd, April 2010.

Kit alone again, June 2011.

Kit with two mares, August 2012.

As well, truly "wild" horses the world over – including the horses in my study area – have possibly mutated, as the wild horses of Namibia did, possessing slightly different DNA from that of domestic equines. The Namibian story runs parallel with that of the horses in the Ghost Forest.

❖ ❖ ❖

The wild horses of Namibia live in the most extreme desert conditions possible, have faced extinction by starvation and lack of water, but continue to survive. There are two stories about their origins. One is that a ship carrying horses from Europe to Australia ran ashore near Namibia's Orange River. Another story links the horses to Baron Hansheinrich von Wolf, who in 1908 lived on the edge of the Namib Desert. His horse-breeding program included thoroughbreds, Trakehners and Cape horses. When the baron died from wounds he received while fighting in the Battle of the Somme in 1916, his wife released three hundred horses into the desert. Likely the wild horses' bloodlines include a combination of both groups of horses that then could also have joined those of other horses released at the end of the First World War. The whereabouts of the Namib's wild horses were kept a secret for more than seventy years, until 1986. In 1992, when the horses seemed to be facing extinction, the government introduced a feeding program during the area's driest months. In 2006 the herds totalled 150 head.

Those who write about the Namibian wild horses, like Sandra Uttridge, say the animals maintain relationships with indigenous wildlife – gemsbok, springbok and ostrich. There is no evidence of competition for the sparse feed that exists, and they can be seen grazing within metres of one another.

Experts internationally attribute their survival in the desert to adaptations in physiology and behaviour. As F.J. van der Merwe writes:

> However, though the horses have a genetic similarity to Arabian-type horses, they do not closely resemble them in outward appearance. Further, in blood typing studies done in the 1990s, a new variant was noted. Its absence from the blood samples of all other horse breeds indicates the presence of a mutation that probably occurred after the horses became established in the desert.

When I read this, I wondered if the same is true for the Ghost horses.

◈ ◈ ◈

In addition to the behaviours and abilities, and possibly some alternate genes, that make the wild horses of the Ghost area different from their domestic, free-roaming and feral cousins, I have also noticed another thing about them: the wild horses I study are natural stewards of the land, using a rotational grazing method suited for this region of the world.

I interviewed a rancher in Grand Valley, Alberta, in early 2012. I asked him what his thoughts were regarding the quibble about wild horses competing with cattle and native wildlife for grass. He was adamant in his reply that of course the wildies eat the browse cattle require when turned out on the grazing leases. He cited an example that he saw as the foundation for his knowledge. He has a quarter section of land with a bunch of horses on it, and they overgraze "platforms," which I gather he meant to be the tastiest grasses. I think that the "platforms" he spoke of are areas of pasture that ranchers see their domestic stock overgrazing, giving a flat appearance to those areas.

In the summer of that same year, Lois and I decided to go on a backcountry horse trip up in the Clearwater drainage of central Alberta. I wanted to see that part of Alberta from horseback, and not having a packhorse anymore, I opted for a more luxurious and restful (or so I imagined) option, that of going with an outfitter. At the same time, Lois was on the lookout to buy a good mountain horse. I had already heard of Wild Deuce Retreats and Outfitting from my friend Pam Asheton years ago as a company that had good horses for sale at their annual mountain horse auction. Pam is the author of *Alberta Backcountry Equestrian One-Day Trail Guide*, as well as someone I admire as a horsewoman and judge of horses.

I spoke to the co-owner of Wild Deuce, Chuck McKinney, who is not only an outfitter but also a horse trainer and cowhand for local ranchers up in the Clearwater / Ram River drainages northwest of Sundre, Alberta. I sought another opinion on the question of whether or not wild horses compete with indigenous wild animals and cattle

Chuck McKinney of Wild Deuce Retreats and Outfitting, through the smoke of his campfire.

The Deer Lake mineral lick in June 2012. Wild horses and other wildlife have been feeding here since I set up the capture camera in February 2012.

for graze. Chuck said, "The wild horses graze the land but have a management sense themselves and are natural stewards of the land. They don't overgraze a platform as a domestic herd would do." He went on to say that in a lifetime of helping neighbouring ranchers work with their cattle on the grazing leases west of Sundre, he has begun to see feral horses on the allotments that were not there before. He did not mean to say that there are more ferals, but rather that the feral horses have migrated, moving in from other areas that have been disturbed by the expansion of recreational motorized vehicles.

To test Chuck's theory about wild horses being natural stewards, I decided to make some observations. At a mineral lick where I have a capture camera, I noted the growth of grasses from February to October of 2012. Six different herds have been photographed using the lick for up to two hours at a time, three or four times a week. During this period, there was no sign of grazing the new clovers and fescues, since they were grazed in the winter. Instead, during that period of time, from spring to fall, the horses were grazing in the uplands, moving high to catch upslope wind to avoid the bugs. Horses that are wild animals know how to rotate their grazing ranges: they return to previous areas with the seasons. They seem to save the marshland for winter; the muskeg is frozen then, allowing safe passage to their "hay," which was not grazed in the wet months.

My domestic horses devour domestic clover in the spring. The wild ones, however, avoid it. There is a lush clover patch in front of one of my capture cameras. With all the horses I see in my photos using the area, I cannot understand why the clover is left untouched. I asked my vet, Alyssa Butters, why this was the case. She told me about alsike clover that damages the liver, causing skin blisters and "scratches" to white-haired areas. I asked the same question of Chuck, and he said some clovers are anticoagulants. Alyssa said, "Clover can indeed cause bleeding disorders, but it is mouldy sweet clover that is the culprit. The toxic substance is dicoumarol, which is formed from weathering of a compound found naturally in the clover. Dicoumarol is typically found in sweet clover hay or silage that is spoiled or improperly cured, seldom in fresh clover." Either way, the wild horses will not eat the clover, fresh or otherwise. I have no idea how

the wild horses know not to eat it while my mare would happily devour it.

Erik Butters admits the theory of wild horses stealing the grasses with the seed heads out from under a cow's nose in springtime originated in part from what he heard his father say. But he and his daughter Erin are rangeland grazing specialists, and both of them cite examples of wild horses competing for their cattle's grass closer to home.

Erin puts it very well. She says horses are harder on grass than cows are, as horses have upper and lower chomping teeth while cows whip their tongues out to grab hold of the grass. So if there is a group of horses on a piece of grassland, they will pose an overgrazing problem. Butters Ranch does not use its allotment to capacity, but Erik and Erin both note it is hardest when "ferals" are out in that area. Both Erin and Alyssa told me they have seen the occasional wild herd on their ten-year renewable lease land. Erik said he has easily chased them out with his quad. At the same time, both sisters agreed that wild horses competing with cattle for grass is a non-issue in the far reaches of their summer grazing allotment. Erin continued to say theirs is an impression based on years of riding, and that range assessments made both close to the ranch and on the far reaches of the allotment, using small fenced plots perhaps, would be a worthwhile endeavour.

On the whole, cattle and wild horses have a positive impact on a wilderness landscape, contrary to the claims of a large majority of wilderness organizations, which are especially worried about damage to riparian areas. On the Butters grazing lease, the fragile riparian areas are fenced off with solar-powered electric fencing, and wild horses certainly don't stand around watering holes during the rainy months of summer. The wildies graze and water in the lowlands, briefly and nocturnally, and return to the windy slopes above during the day.

However, perhaps the greatest value of both cattle and wild horses is their control of brush. Without their passing by on a regular basis, the forest north of my studio would be difficult to ride a horse or walk through. They clear the understorey of low dead branches and brush, thereby creating more biodiversity. Their grazing of the lush grass reduces the chances of accidental ignition when the grass turns dry later in the year. In addition, wild horses literally fertilize certain areas of forest, increasing forest strength.

On the coast of British Columbia, I have seen places where grizzlies and black bears year after year haul out the salmon they catch to eat in peace under a cedar tree. The cumulative impact of a bear's salmon scraps and scat cause the trees in the vicinity to be larger than those nearby. There is scientific documentation that some of the growth of the large cedars along the West Coast is the result of numerous generations of grizzlies feeding under them.

Like the bears on the West Coast, wild horses go to certain trees year after year, but for different reasons. Almost ritualistically, they choose a tree to rest near and avoid the summer sun and insects. While there, the herd fertilizes the large spruce and pines of the forest. These, like the cedars chosen by bears, have a tendency to be largest in the area.

I wrote to Dr. Claudia Notzke, a respected wild-horse researcher and associate professor at the University of Lethbridge, to find out if she thought wild horses have a positive impact on forest understorey management. She replied: "As to horses being used as 'ecosystem engineers,' in Europe there are quite a number of projects where 'primitive' horse breeds/subspecies are being used for biodiversity enhancement or restoration ecology, such as the Hutewald Project in Germany (using Exmoor ponies)."

◈ ◈ ◈

So wild horses all over do seem to have similarities in behaviour, in ability to live in a wild environment and not exploit it. It would seem that truly wild horses likely do not constitute a threat to cattle ranching. However, there is a stronghold of ranchers out there who maintain that the wildies are not "valuable," that they do damage the pursuits of ranchers and that they certainly are not able to govern themselves sustainably. Such people would likely continue to suggest that wildies spread disease and that their small numbers can only mean that they end up being inbred.

Given the proximity of the wild horses to their domestic horses and those on Native land, I asked Alyssa what her thoughts were on wild horses carrying infectious equine

I asked Erik what he thought of the wild horses I had studied. He answered while I took his portrait: "When you see any wildlife in a perilous environment, you have to be respectful of that."

diseases or parasites. She said Alberta winters are pretty good at killing off the burden of parasites on the ground. The wildies' extensive grazing patterns cut down on their parasite loads as well, unlike most domestic horses, which spend their lives in paddocks. As for diseases, Alyssa did not think there was anything to worry about, with the exception of the wildies being susceptible to West Nile via the mosquito (*Culex tarsalis*) that hatches and transmits the virus. There hadn't been a case of West Nile in Alberta in recent years, but two cases were reported in 2012. The threat of that virus to humans and other animals seems to depend on weather: how many of the right species of mosquito hatch and survive.

Of course, there are other diseases to which wild horses would be susceptible, such as tetanus (a soil-borne bacterium, so not contagious horse to horse) or respiratory viruses. The risk of transmission to domesticated horses would seem pretty small.

As to the speculation that wild horses are only bound to become inbred and weak, wild-horse behaviour seems to guard against this. The herds' populations seem to be relatively stable, with four-year-old stallions joining bachelor herds and fillies allowed on the fringes to be claimed by bachelors. The "rewilded" horses seem to regulate themselves to avoid line breeding (a stud breeding his offspring). Their numbers seem to be controlled for the most part by the cull of the wild, which would include natural predators such as cougars and possibly wolves and grizzlies, not to mention bad weather. There does not

seem to be a weak link in the horses caused by inbreeding. Amongst the wild herds of Sable Island and Namibia, if a weak horse is born to the wild due to a bad genetic link from any source, the animal usually dies. The only weak link from my perspective appears to be that posed by human interlopers upsetting the balance of the ecosystem where the horses live.

Instead, I believe we could learn a thing or two from wild horses and the ways in which they manage themselves – in fact, a few smart horse people have already done so.

◇ ◇ ◇

I always viewed Monty Roberts as the man who created the "tipping point" relative to changing beliefs in North America and abroad about how to start colts. He gained his insights from watching a lead mare school a badly behaved colt in her herd.

While riding up in the Clearwater area with outfitters Terri and Chuck McKinney, I was stunned to discover their passion for the wild herds there and how they modelled some of the management of their summer dude string after the behaviour of lead stallions and mares. Monty Roberts was one of the first to write about his source of gaining trust with young horses, and I am sure other trainers have used techniques gained from watching wild herds, but Chuck and Terri's actions in this regard constituted the first example I was witness to.

The McKinneys' methodology seems simple in principle

but more complex than I have space to relate in this book. Simply put, Terri explained how, in every wild herd, there is an alpha female and male – the lead mare and the stallion that control their herd with barely a sidelong glance. Chuck and Terri have worked out how to manipulate the interaction within their summer riding horses to assure that they (Chuck and Terri) are in the alpha positions. I noticed that the horses kept an eye on both Chuck and Terri when they were in the horses' vicinity. For example, one day eleven of us were riding on a high, wide plateau overlooking the Clearwater River, and we dismounted for lunch. I was expecting all the horses to be hobbled, given there were no trees to secure them. However, Terri and Chuck's wranglers came by and secured the bridles to the saddle horns, letting all the horses loose to graze. We were up there for over an hour, and not one horse strayed. They stayed near the herd bosses, the alpha male and female, Chuck and Terri, who spent their lunch hour nonchalantly chatting to various guests.

To further illustrate the theory that wild horses are doomed to be inbred, losing quality and health, I give you the example of Mooch, who proves the opposite to be true. Terri McKinney had a domestic mare bred by a wild stallion. The resulting foal, Mooch, became one of Terri's favourite horses. He stood fourteen hands, two inches tall (about one and a half metres) and was strong and sure-footed running through timber. Terri described how he was big-boned, had solid feet and was inclined to be an alpha horse although gelded. Once when she was

Terri McKinney chats with me while her stud stands calmly amongst their free-grazing guest horses.

riding him at the front of her line of guests, she noticed that he kept glancing toward the rear of the line. All the other horses were oblivious to the cougar that was walking not far behind the line. Her half-wildy's instincts for survival in the wild were intact even though he was born and trained domestic.

And there are other examples of stories told by those in the know, of how wildy genes are not weak but sought after. At the celebratory launch of the Wild and Free website educational initiative at the Calgary Stampede, a respected horseman of the Stoney Nakoda Nation and his wife sat beside me. Images of the Horse Lake wildies on video footage collected with remote cameras for the initiative were being screened. He laughed and said he knew where those horses were located. He proudly continued, saying he trapped horses in that area, taking the mares to cross with Stoney stock in order to upgrade his herd.

Hank Snow, a member of band council for the Stoney Nakoda Nation, who once laughed about the white man's assumption that horses arrived in North America with Columbus, also talked about the value of wild horses to the Stoney people. On the Galileo website, he is quoted as saying wildies are "tough mountain horses, can cross the swamps easily and survive on little."

I interviewed Hank Snow in the early 1990s while working on *Grizzly Kingdom*. As a guide, he hoped to show me a grizzly on Stoney land below Mount Yamnuska, the first mountain to the northwest visible from the Trans-Canada Highway heading toward Banff. We did not find a grizzly but did see wild horses of the type he described. They galloped off, leaping rock escarpments, racing upward toward the shoulder of Yamnuska without a stumble. It's no wonder these horses are sacred to the Stoney Nakoda people, as I was about to learn.

❖ ❖ ❖

The issue of sacred sites formed part of my attempts to bring some protection to the southeast corner of the Ghost Forest. I had asked Peter Snow, environmental planner and consultant for the Stoney Tribal Administration, if there were any sites in the area, because if so, they would have to be identified before logging could take place. He replied that no surveys had been conducted, a point I brought forward to SRD, who brushed off my inquiry. But it was Snow's response about my having found sacred sites that surprised me: he said the wild horses themselves are sacred to the Stoney people. I was seeing them constantly!

I recall Eleanor O'Hanlon writing to me, thinking of Lakota holy man Black Elk's great vision, with the horse as the incarnation of creative power: "He [the stallion] dashed to the west and neighed, and horses without number, shiny black, came plunging from the dust. Then he dashed toward the north and neighed, and the east and to the south, and the dusk clouds answered, giving forth their plunging horses without number – whites and sorrels and buckskins, rejoicing in their fleetness and their strength." Magnificent.

❖ ❖ ❖

The wild horses are on alert when they hear a quad or bike approach, as many do on paths not designated for ATV use. Most of the time the horses move up into the forest. They could become habituated to the noise, but why should they? They need relative quiet in order to survive.

As I conducted my study of the horses and saw the effects of ATV use on them and the other animals in the area, I wondered if I were overly concerned about ATV noise. But Erin Butters also voices great concern, telling me that the Butters' open-range beef cows respond negatively to the noise, that they put on less weight when they are subjected to anything that causes stress; vehicular noise emerging suddenly out of the quiet forest is a concern. Cattle grazing near road allowances are habituated to the noise, not so the creatures of the forest and free-roaming cattle of the grazing lease. I further wondered about the impact of the expanded ATV network in Alberta on the remote wild-horse herds (not the ferals by the roads, as they would respond as habituated cattle do).

When I rode with Chuck and Terri McKinney – observers extraordinaire of wild-horse behaviour – I asked Chuck if he had seen an increase in the wild herds in their area, as I had heard Erik Butters talk of more stud piles closer to the home section of the ranch than ever before. Chuck's response was interesting. He said he had heard a lot of talk from SRD about the numbers increasing, but he had not seen it. What he had seen were wild horses in the Radiant Creek and Falls Creek areas eating the new growth of grass, thereby depleting the range when the cattle come out in mid-June. He thought the horses came from east of the Ram River, in some very remote areas north of Highway 40, as he recognized a couple of the herds. He said he suspected they moved south because of the ATV noise. Wild horses, like any wild animal, shun noise. Chuck's suspicion is consistent with what Erik said about seeing more feral horses near his ranch: the Butters grazing allotment was opened to ATV use in 2006.

The problem is not really the ATV folks, but rather the design of the location of their permitted-use trails. The ATV trail system in the southeast corner of the Ghost Forest is dysfunctional from a watershed perspective; in fact, its design encourages illegal trail-breaking. There are many dead ends on the permitted-use trails in the area. Who would want to turn around and return the way they came? The ATVers understandably create their own loops to avoid the poorly designed trail system. The loud noise made by machines tearing through mud in the muskeg is offensive to my ears. I can only imagine how a wild animal would feel.

One of my capture cameras revealed an interesting sequence of events on Camp Ridge which clearly documented the ridge being used as a wildlife corridor between the Horse Lake marshland and the Owl Creek valley beyond. About once a month a group of dirt bikers fly over this same ridge, creating an illegal link between the dead-ended SRD-designed trail system. Once a month

Wild herd on Camp Ridge.

Small black bear walking the trail as wildies and wolves did an hour before.

Bull elk on the Camp Ridge game corridor shared with wild horses, bear and wolves.

does not seem to cause a problem, as evident by the wildlife using it, but if increased, it could be problematic and cause shifts in game and wild-horse movement.

When working on the Wild and Free project, I encouraged an elementary-school art teacher to have her class explore ATV use in the district, as she is a quad-in-the-mud fan. Her descriptions of the latest slip, splash and slide through the forest almost motivated me to try riding an ATV! She also mentioned how sick she was of all the "greenies" dumping on quad users. She said she wished the government would do its job and set aside areas where ATV users are not damaging a fragile ecosystem, and provide the law enforcement to back it up.

One of three dirt bikers roars across the Camp Ridge game corridor, far from designated legal-use trails.

Dirt bikers carve their way through the once pristine pine forest, causing erosion and noise that disturbs wild animals with their passage.

Rod Green filming and talking about the damage to the spring.

Wolves using the same game trail as wild horses on Camp Ridge. Wild Horse Lake is below.

In the first years of my wild-horse study, I called the Sustainable Resource Development offices in Calgary to report ATVs sighted tearing up the marshland. Knowing their response ahead of time became a joke – albeit a painful one. The nice ladies on the other end of the line always laughed politely, "Staff is being reduced, not increased." They always said they would make note of my report. I soon realized I was wasting my time.

My friend and art dealer Rod Green described his discovery of the mutilation of a flowing freshwater spring in June 2012, just before we saw a bachelor herd near Deer Lake. The damage was on an ATV track far away from designated trails: "We are standing on a quad track approaching a sensitive marsh area and the destruction is getting worse. The quads have torn a strip through a spring leading into the marsh. Weekend warriors are having a good time at the expense of the horses, wolves, bears, moose, foxes and sandhill cranes that populate the area. Too little an area, too small a problem and no one gives a damn!"

Logging in the Ghost Forest is also a potential problem for the ecosystem there. I asked a friend of ours, Marina Krainer, executive director of the Ghost Watershed Alliance Society, what knowledge she had of Spray Lake Sawmills' (SLS) designs on the southeast corner of the Ghost Forest. In my travels through the area, I have seen cutblocks marked with pink ribbons and striped black-and-pink ribbons ominously indicating the planned road across marshland and up over hills. I spoke to an employee of SLS who was tying on the ribbons. He said the road would come in along the existing road that leads through the Bar C Ranch and into the southeast corner. However, the main road they will use will come in from the north. I am not opposed to logging – after all, I live in a wooden house built from local timber. If SLS cleans up their slash, as I have noticed other companies doing farther up Highway 40, logging in the area would be good for cattle, the wild ungulates and the horses. If done correctly, the landscape could remain beautiful. The larger problem is the roads loggers build and do not reclaim, thereby opening the land for further illegal use of landscape by ATV users and others.

The folks in this track vehicle illegally crossing Deer Lake marsh in 2007 said they were not harming the wetland – but they paved the way for future ATVs to cross on a non-designated track.

The track vehicle in the photo opposite set the tone for this quad recreational users off-limits area. We found a ruddy turnstone nest with four brown eggs right beside one of the tracks.

Marina is adamant that the damage done is unfortunately not surprising. No enforceable solution is in sight given the badly designed ATV trail system in the area and the fact that there is no one out there to oversee legal trail use and catch the widespread illegal use.

In the end, I believe that the problem of protecting the wild horses and their fellow species in the southeast corner of the Ghost Forest will have to be solved by the ATV users, the loggers and the government that designs and allows for their access to the area. Not all of them are bad and irresponsible. They are fully capable of designing the solution and being part of the implementation of the area's protection. I don't think the solution is law enforcement, because there could never be enough. Areas where ATV use is appropriate need to be established and other areas need to be closed completely.

The value of the wild horses and their cohabitants should not be in dispute, and we can work together to solve the problem. I worry that if we don't, we will be revisited by the Stoney man, Jacob Swampi, whose spirit has been known to avenge the exploited environment.

◈ ◈ ◈

In 1939 and into 1940, Stoneys and non-Aboriginals alike in the Ghost Valley talked about Jacob Swampi's ghost. Erik Butters told me the story of Swampi's revenge on July 24, 2012:

In the early 1900s, Spray Lake Sawmills was logging up there. They made a deal to cross Butters Ranch, cross Indian land and into the Forestry. They hauled out, over three winters, 5,600 semi loads of logs. They all stopped at the ranch before they turned onto the highway, to make sure everything was tied down, nothing sticking out or anything like that. My buddy Keith Mumford and I were out feeding cows. This guy jumps out of the truck raving, "I don't know what's going on around here! Do you know what's going on around here? Things are crazy around here." I asked what he meant. "Up at mile four up on the Indian Reserve, this guy jumps out in front of the truck in the middle of the night – dressed in a buffalo robe – waving his arms." This went on for a week or so with different drivers. The sawmill folks called the RCMP, but they wouldn't even go beyond the turnoff into Mom's log home.

In the meantime, a neighbour, Keith McMillan, who was a general investigation officer with the RCMP in Calgary, phoned me. Officer McMillan related that he had heard something wacko about an old Indian in a buffalo coat and a rifle spooking the truck drivers up my way and asked what I knew about it. That same afternoon the trucking boss came along and asked what was going on and who was pulling that stunt. I told him I thought it was

Jacob Swampi. The boss asked: "Where do I find this guy?"

I said, "It will be a little tough, as he has been dead since 1939."

The foreman got quite huffy and left.

Officer McMillan called me again and said he thought that was prime! Word got around about not being able to find any tracks of this guy in the buffalo coat. The loader up in the Forestry on the night shift had about a twenty-minute wait between trucks. This guy must have been about four hundred pounds, but he heard the story about Swampi's ghost being the guy potentially causing the grief. He wouldn't stay up there by himself, so he would load a truck, jump in with the driver and return with the empty truck going back. This went on for a week or so. Then it stopped.

Folks thought it was someone playing a joke, not wanting logging. Everyone who it could have been was accounted for, except for Jacob Swampi's ghost.

LOIS EMERGES FROM HORSE LAKE MUCK WITH *ZENITH* BRONZE TABLET.

CHAPTER SEVEN

Zenith: What the Cameras and the Art Reveal

❋

I am not sure what drew me to the project. Perhaps it was the joy of riding deep into the forest with Maureen to try to catch a glimpse of something so ethereal. It was as if we were a part of a special secret that was just ours. Nothing compared to the beauty of the aspen forest or the wolves howling through the valley.

It was while walking through the aspens that I began to notice an interesting play of light. As I glanced through the trees on a bright day, I saw no horizon – just the contrast of light and dark. The "wildies" in my tablets are ghost images, floating in and out of the shadows.

The title of my piece in the photograph is Zenith. The number of horses that have gone back to being wild animals is declining in Alberta and indeed in many other places in the world. It is my sincere wish that this not be their highest moment, but that they enjoy resurgence and earn a respected place in our hearts and minds. They have, dear reader, captured mine forever.

—LOIS GREEN

◈ ◈ ◈

I began my study of the Ghost horses and those that share their ecosystem by riding Hope and hiking into the area armed with long-lens and point-and-shoot cameras carried in my saddlebags or backpack. Since I installed the four capture cameras in 2012, my thoughts about the rewilded horses have taken a spin into a world I had not imagined possible. Most surprising are the events recorded in the dark of night. More than 75 per cent of images taken by the cameras from early July to the end of September are nocturnal, from cameras on the valley floor. The opposite is true of a camera on a ridge about 250 metres above Horse Lake. A capture camera has a limited field of view from left to right but provides a great view into the "beyond" straight ahead of its lens. As I made gathering the cameras' images a practice, I began to see the field of view as a canvas, or a stage. During the summer months of 2012, I excitedly downloaded images every week or two, and that activity has been better for me than any visit to an art gallery or theatre.

◈ ◈ ◈

A curtain of snow caught with a single horse.

Small herd with a yearling in February, Deer Lake.

A wild horse's eyeball becomes a transparent orb.

Deer in the dark of night cross the stage as a moon attempts to illuminate the scene.

I do not think I am atypical in feeling that if I can't see it, not much is going on. I have three heat- and motion-activated cameras installed near mineral licks and vantage points I have learned the horses use frequently. The first chip I downloaded into my computer was from the camera installed at the end of February from the Deer Lake lick. One ghostly apparition after another was captured of three different horse herds at night. Deer entered the field of view with about the same regularity as the horses, often seconds apart.

I gasped with astonishment seeing two timber wolves caught breeding within the frame of the camera. Over a two-minute time span at 7:08 p.m. and 7:09 p.m. on February 24, 2012, with the last sliver of moon in the sky, a black wolf courted and bred a light-coloured female. She even looked at the camera.

Horses that seem gigantic when only centimetres away from the camera are small when silhouetted against the last light of day at 7 p.m., the Rocky Mountains captured surrounding the contours of their black shapes. As night progresses in the last gasps of an Alberta winter, a herd huddles for warmth in the dark: ghostly apparitions, the infrared shifting their dark colorations to light. An eyeball, inches from the camera, becomes transparent, revealing animals beyond its orb. Horses dark by day become white by night, the revered white horses of the Stoney Nakoda Nation revealed. Their hides, so close I feel I can touch them, show every bite and scratch of their latest fights.

I gained an intimate relationship with these animals as they stood a few feet from my lens, which allowed me to

Wolves in late February prior to breeding at Deer Lake mineral lick.

A small wild herd warming one another as the sun goes down.

A lone wolf visits the mineral lick as a meteorite streaks across the distant sky.

Bull moose testing the conception level of a cow in spring.

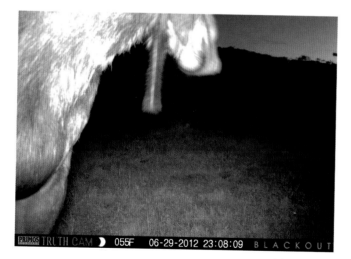

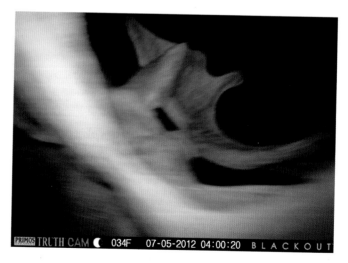

A bull moose so close at night one could reach out and pull his dewlap.

Bull moose antlers in the night vision of the capture camera.

look at their world through their eyes, as I had done a decade before with three grizzly cubs in Kamchatka.

As dawn approaches, the horses' world takes on a bluish tinge in the camera's eye – a single stud, Kit, is immobile on frozen ground with the Devil's Head beyond. The snows of April reveal the reverse prints of many animals as a horse shifts position, its body looking as if captured by X-ray. Small, bright white orbs of eyes in the black beyond are caught by the red-eye flash.

Moose and cougar wander across the stage, defined by the camera's window in the moments after midnight in April. A wolf's fur seems as if it is constantly being updated by a fresh application of dark hair dye, unaltered by hours of bleaching sun; the wolf trots across the frozen ground at 6:00 on a morning in May, the Rockies caught in ultramarine blue, with the kind of light only winter's end can produce. At 9:30 p.m. Kit seems to pose, light gleaming off his eyeball, as the sun sets over the distant Rockies. On June 5 at 5:26 a.m., spent dandelion heads catch the light as alpenglow turns the Rocky Mountains red and fog hovers over the marsh, chilled by the night air. Just before midnight on July 13, a young bull moose tests the reception of a young cow to his amorous advances. She indicates it is too early in the season by laying back her ears.

From May through August, the bachelors remain in the Deer Lake vicinity. They are chased away by stallions with mares coming into breeding mode. At Horse Lake, over two kilometres away, mares with new foals, stallions on

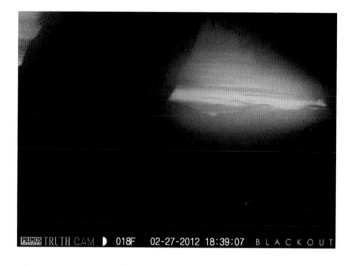

The Canadian Rocky Mountains are framed within the shape of a wild horse inches from a capture camera.

Stud in the marsh on an early spring afternoon.

A wolf with full winter fur trots across the frozen ground in early May.

Bachelors spar in early light that glances off the spent heads of dandelions.

A new foal catches a night snack.

A foal stands within the protection of its dam and a young filly.

alert, all move in and out of the frame with a black wolf seemingly on guard. At 2 a.m. on May 30, a newborn foal stands within the envelope of its dam and a young filly. On June 1, in the light of a waning moon, a stallion guards his beautiful filly. At 4:25 a.m. on July 3, a cow moose with twin calves strolls across Horse Lake in the light of the full moon, and a lone stallion leaves the camera's defined stage.

◈ ◈ ◈

"As shown by your study, Maureen, I think an 'Aha' moment has occurred in the Ghost Forest. What humans will do with the information will mark mankind as to what we will become."

— ANITA CROWSHOE

Moose enter the established canvas of the capture camera in the lowlands. A bull, cow and calf all participate. The cow poses with her body so close to the camera that she fills the stage, her hide inviting the viewer to touch it. She and a bull stand to frame a view of the bachelor herd at the edge of the marsh. Another cow appears with her calf suckling in a vision of contentment before the pair gaze in unison at an approaching wild horse. The bachelors run toward and away from the camera as if to show off their perfect oval-shaped hooves, which have never been seen by a farrier. A wolf pup now four months old trots into sight, pauses briefly and leaves.

Higher up, the camera positioned on Camp Ridge richly reveals the entire region of wild-horse haven. The lake, with early mist rising off its southwest shore, is framed by El Cid's herd frolicking in the first sun of the day, their hides a patina of ebony and copper.

To my total surprise, over a few days in August, the whole menagerie of the wild strolled by, with Horse Lake below and the Rocky Mountains beyond. First, two bull elk pose as if framing this special place before a wild-horse herd lingers, their backsides blending with the curve of the uppermost part of the ridge. A day later, a lovely, dark-cinnamon black bear pauses before walking on, followed a few days later by two wolves. A fox and later a coyote, seemingly not wanting to be left out, trot across the camera's view, followed shortly thereafter by another wolf.

I say to myself that some kind of bizarre harmony is going on here. I laugh my head off at the antics of the deer peering into my camera or holding an ear in front as if to intentionally show off its ability to gather any sound of approaching predators. When the bull elk puts his nose up against the camera, his long legs framing the bit of country that is still so close to pristine and a horse herd beside the lake below, I cry.

◈ ◈ ◈

Rick loves examining scat more than I do. In September 2012 we hiked to the top of one of the ridges to change batteries and film chips we had located near a strategic mineral lick overlooking the entire Horse Lake and environs. En route, Rick picked up a fresh bear scat, astonished to see it was full of dried spruce needles. Neither of us could imagine seeing a black or a grizzly eating spruce trees, dead or alive. When we found the scat, we were near the hill that overlooks Horse Lake, which we call Camp Ridge. A nice creek runs out at its base. The meadow merges into dense spruce and aspen forest, with game trails running in all directions indented by horse, wolf, deer and moose tracks. We saw anthills, damaged by a black bear seeking a meal of ants – high in protein, and a delicacy in many parts of the world on the human menu too. Many spruce needles mixed with other duff lay scattered about the former ant homes. The loonies dropped for both of us at the same time. The black bear was feeding on ants and it ingested and passed the spruce needles through as a result. Duh! That night, a large black bear dominated the screen where minutes before a herd of wildies had been at a mineral lick.

El Cid's herd moves up and down the steep hill and, moments later, another black bear sniffs at their tracks. A moose also poses overlooking Horse Lake far below. Rick, upon seeing that image, exclaimed, "What is a moose doing up here?" All we were missing was a grizzly bear, but it was September 8, 2012, and the kinnikinnick berries were ripening to cover Camp Ridge. I thought to myself, "Grizzly bear and cub that I saw by the Owl Creek herd twice this year, it is your turn to complete the story."

Mule deer eyeballs the camera.

Buck near Deer Lake.

Cow moose and her calf at Deer lake mineral lick.

Cow moose feeds her calf.

Bachelor stallion in front of capture camera.

Bachelors at the Deer Lake mineral lick.

Fox.

Young wolf rolls in front of capture camera.

Bachelor stallions play-fight.

Wolf stares at the capture camera.

Lone stud watches wolves pass by.

Stud and his filly.

Cuero's herd, 2012.

Wolf examines unidentifiable critter.

Morning at Horse Lake.

Evening at Horse Lake.

Herd above Horse Lake.

Bull elk bugling along the Camp Ridge game trail.

＊ ＊ ＊

"It's the little things in the environment that matter most. Krill and phytoplankton in the ocean, honeybees on the land, and small, sensitive land areas. The Ghost River Wilderness is such an entity that matters, supporting a rich and diverse zone of biological communities."

– ROD GREEN

Valley of the Ghost

Sacred land, endless sky
Call of the wild, hear me cry
Take my earth, take my star

Devil's Head, spirit rock
Shifting shapes, through the dark
Jagged path, torn apart
Twisted trail, broken heart

(Chorus)
Give me freedom, give me hope
Give me land, no rope
I am free, I am speed
I am strength, I can bleed
I am home
In the Valley of the Ghost

Rhythm runs, through the beat
Of the race, flying heat
Never sleep, keep up the pace

Falling steps, through the trees
Storm clouds, lightning breeze
I am your image running out of time
Broken trail, broken heart

Where horses still run free
Phantom warriors proud steed
Souls of spirits, carried by the wind
Place of refuge, closing in

© CORI BREWSTER, JOHN CAPEK, RICK PRICE

＊ ＊ ＊

He showed my mare Hope and me the silence of the forest as deer do when they hide. He revealed the silent language of equus in 2006. I thought Cuero had disappeared, perhaps having been trapped and killed, because he and his herd seemed to disappear for three years. It wasn't until the fourth year of my study, on the last day of October 2011, that I saw him again. I excitedly almost rode my horse over an embankment. Posed on a ridge beyond was Cuero, backlit by the setting sun, a shape distinctively his.

Hope and I ritually look at the wild-horse landscape before I ride down off Salter Ridge to my truck. In May of 2012 Cuero was on Camp Ridge once more, standing uphill from his herd to guard the newborn foal with the chestnut mare, as well as others of his family. Later, as if knowing I own the hidden camera, he nonchalantly walks up to my lens. Moments later, a wolf follows, the two peering one after the other into the lens. I imagine

Cori, Camp Ridge, with wildies below.

him now with mares and foals, caught as if in an impossible limbo, peacefully walking the mud flats of Horse Lake with a pack of wolves. This is a horse that has become a wild animal, having formed an alliance with some wolves for his survival, sharing his range in peace with the deer, the moose and the sandhill cranes.

I wonder what this ongoing study of wild horses has done for me as an artist. It all started with Cuero and the awe I felt the first time I encountered the bachelor herd, and it has ended with me coming to grips with the relationships I've seen between members of the Ghost community: horses, wolves, deer, moose, cranes.

Wild horses and wolves fascinate me. The wildies represent the freedom that is vital for my own emotional and intellectual survival. The wolves hold a fascination rooted in vital power. The two combined are omnipotent. I realize, as I conclude my thoughts and observations, that wild horses and wolves represent who I am as a visual artist.

The drawings began in awestruck charcoal black-and-white interpretations of how I felt when I met Cuero and the day Hope responded to the silent language of equus. My first paintings reflected the natural beauty of wild horses in their almost pristine marshland habitat. As I began to realize there was more to the wild herds than running free, hints of my understanding of their critical relationships with wolves and other wild animals emerged. These paintings were executed in bright colour and bold lines that merge to express intuited understanding. My final paintings are more abstract, expressing what I believe to be the case – but am not certain – about the horse / wolf symbiotic relationship.

If wolves and other wild animals can form critical alliances with wild horses for their survival, if they can accept the rewilded creatures into their indigenous midst, perhaps recreational users, government officials, commercial users and hunters can accept the horses' importance too.

➤ CUERO AND HERD WATCH HOPE AND ME. ⬅

AFTERWORD

Protection of Wild Horses in Canada and the United States

⁜

Conservation of wild horses in Canada and the US is complex. However, this much is clear for me: whatever policies we draw up and enforce relative to the horses, I do feel strongly that we as human beings need to allow the truly wild herds enough isolated wild land to run free.

After reviewing the mixed bag of approaches to protecting wild horses in Canada, I have concluded that the rewilded horse, such as is found in pockets of the Nemiah Valley in BC, parts of Alberta (including my study area) and Sable Island, needs federal protection. The protection of wild horses in the US, which is operated federally under the Bureau of Land Management, is under review, with changes on the horizon. I will briefly summarize where some if any protection exists in Canada, and even more briefly the admirable work of Velma Johnston and the protection of wildies for which she is credited in the US.

The Nemiah Valley Wild Horses

The Nemiah Valley Wild Horses are located in the Chilcotin interior of British Columbia between the Chilko and Taseko Rivers. In 2007 the Supreme Court of BC found that the Tsilhqot'in people had an Aboriginal right to manage eight hundred to one thousand free-roaming horses for their own use. The Xeni Gwet'in First Nations Government established the Elegesi Qayus Wild Horse Preserve in 2002. In her 2012 unpublished sabbatical report on wild/feral horses on protected lands in Canada, Dr. Claudia Notzke writes:

> The Tsilhqot'in people of Xeni comprise one of six Tsilhqot'in First Nations groups, each of which has an elected local government. The Xeni Gwet'in have always been very proactive in asserting management responsibility in their ancestral lands, and recently were partially successful in a major court case for rights and title. They are developing an ecosystem-based forest management model, an Access Management Plan and eco-tourism initiatives that will include cultural experiences and wildlife viewing. They have declared much of their territory a protected area, designating it the Aboriginal Wilderness Preserve (1989) and the Elegesi Qayus Wild Horse Preserve (2002). Wild horses have

always taken centre stage in the advancement of their resource and land-related interests.

The Friends of the Nemiah Valley, a grassroots conservation organization, funds a ranger program for the horses' protection within the preserve of the Xeni Gwet'in. The province of British Columbia still refuses to recognize the wild horses there as a species with a right to remain on the land, even though historical evidence suggests that the horses long preceded European settlement. Friends of the Nemiah Valley will continue to argue for their recognition as legitimate wildlife.

The Wild Ponies of the Bronson Forest, Saskatchewan

A law was passed in 2009 by Saskatchewan's legislature to protect the wild ponies of the Bronson Forest, 170 kilometres north of Lloydminster. The ponies are few in number, but with one of the clauses in Bill 606 reading "No person shall in any way willfully molest, interfere with, hurt, capture or kill any of the wild ponies of the Bronson Forest," perhaps their numbers will increase.

The Sable Island Ponies, Nova Scotia's Coast

Dr. Notzke has permitted me to include her current and apt assessment of the Sable Island ponies here:

> Sable Island was already a Canadian landmark when discussions were initiated in 2010 to have the island designated a national park, thus providing the strongest possible mandate for the protection of its natural environment. However, this prospect also immediately called into question the future fate of Sable's wild horses, which have called this island home for over 260 years....
>
> Critics have suggested an alternative designation for Sable Island as a National Landmark or a National Historic Site to permit the continued existence of its equine inhabitants while avoiding a conflict with the law and policies which are meant to guide national park management. Nevertheless, after extensive public consultation Parks Canada made its position on the topic of Sable Island's horses clear: The horses have been living on the Island since the mid-1700s, and are therefore considered to be part of the ecosystem of the island. All *wildlife* on the island, *including horses* and birds, will be protected under the Canada National Parks Act. ...
>
> The consultation process revealed a broad-based public concern that the Sable Island horses be protected, and be allowed to remain wild with no active human management. Another recurring theme was the recognition of the cultural significance of the island for Nova Scotians. ... Parks Canada is clearly trying to avoid a public outcry like the one that followed the government's first attempt to remove the horses prior to 1961. With its declaration of the horses as wildlife it acknowledges them as a naturalized species in the island ecosystem, an

ecosystem which has continued to thrive and includes the largest breeding colony of grey seals in the world, an amazing avian fauna (including the endemic Ipswich sparrow and the roseate tern, both species at risk) and over 190 different plant species. This action also constitutes a tacit admission that there is not always a sharp dividing line between natural and cultural heritage. Not all cultural heritage is man-made and inanimate.

Thanks to concern exhibited by many Canadians such as Dr. Claudia Notzke, Sable Island National Park Reserve is Canada's newest national park, the 43rd to be created in Canada's family of national parks.

Virtually No Protection of Wild Horses in Alberta

Environment and Sustainable Resource Development (ESRD) is the environmental management branch of the Alberta government that oversees the so-called protection of the free-roaming "feral" horses in Alberta. They have established the guidelines for a Feral Horse Capture Program, which is detailed on the ministry's website.

However, ESRD manages the "protection" of wild horses with an eye toward their long-term survival and the "protection" of public and natural resources with a gross misunderstanding of both the horses that have rewilded and the ferals. It is a no man's land east of the Rocky Mountains, with minimal law enforcement for ATV trail use and virtually no policing of the capturing of wild and feral horses.

The ESRD guidelines for feral horse capture state:

All captured horses can be inspected to ensure that they are feral horses (i.e., no brands) and that the licence holder has adhered to the conditions of capture. Livestock Identification Services Inspectors are contacted if any branded horses are captured. Once inspected and the animals are removed, it is difficult to know for certain where they end up. Most are used by the licence holders for resale, as packhorses, or as rodeo stock. Some are domesticated for various recreational pursuits.

I have heard from locals, however, that feral horses are rounded up with no inspection occurring and that for the most part they are shipped to horsemeat plants at Fort Macleod or Lacombe.

ESRD minister Diana McQueen, in a November 13, 2012, letter addressed to my partner, Rick Kunelius, wrote:

I am happy to tell you that departmental staff recently met with members of the Wild Horses of Alberta Society to discuss the society's concerns about free-roaming horses.

Minister McQueen's letter went on to say that ESRD is attempting to develop a management approach for free-roaming horse populations in the Eastern Slopes, but that as of November 2012 there had been no decision on whether licences will be issued for the upcoming capture season.

In recent correspondence with Bob Henderson of the Wild Horses of Alberta Society I was informed that no licences for capture of free-roaming horses are to be issued in 2013.

Now I am not opposed to culling the herds when necessary, or even to the slaughter of horses for meat. But I am appalled at the uninformed and unpoliced process of horse culling in Alberta. How can ESRD know what horses need to be culled with no scientific studies made relative to the ecosystem in which the horses live, including both wild landscapes and cattle grazers' allotments? Of course, I know my study area the best, but I know that the Ghost Forest serves as a microcosm of the kinds of problems faced by wild horses throughout Alberta.

Managed as a Forest Land Use Zone (FLUZ), the Ghost stretches along the foothills of central Alberta approximately from the Red Deer River to the Stoney Indian Reserve. It borders parts of Banff National Park to the west and follows a trajectory north of the Rabbit Lake Indian Reserve on the east. My study took place in the southeast corner of the FLUZ, adjacent to the Rabbit Lake Reserve, east of Highway 40. The wild horses there have been relatively protected by their geographical location; by Butters Ranch, which manages the Aura Cache Permit; and by the foresight of the local horse trapper, who did not capture all of them.

Unfortunately, with possible expansion of ATV use after the area is opened up with logging roads, the once-pristine wilderness could disappear. The interactions between the wild horses and the wolves and other species in the area would no longer occur. I originally felt that this ecosystem should remain the secret it has been for almost a hundred years. But I've changed my mind. If people are not informed, they will not know what they are losing as a result of bad management on the part of the government.

This book may speed up the demise of the horses that have become wild animals, or – and I hope this is the case – it may provide the catalyst for change necessary to conserve the magic that is happening amongst the herds, wolves and other wild animals. The wild horses of the southeast corner of the Ghost FLUZ contribute to the delicate balance of natural forces that makes up their home. The marshland of the wild horses deserves special care and attention.

Canadian sales of ATVs have tripled in the past decade, having reached $1.08-billion in 2007. The Canadian Off-Highway Vehicle Distributors Council says 2011 sales (including motorcycles, scooters and parts as well) totalled $1.268-billion. Canada is now the largest ATV market per capita in the world, and Alberta alone accounts for fully a quarter of that. As mentioned previously, there are some good people in the ATV crowd, people who care about Alberta's environment. These folks, as well as the companies interested in logging the southeast corner of the Ghost FLUZ, need to be part of a plan to conserve the area and the rewilded horses of my study area on the Butters Ranch grazing allotment.

South of Alberta's Border

Velma Johnston, better known as Wild Horse Annie, grew up in Nevada and witnessed atrocities both in capture methods and transport of wild horses to slaughterhouses. The letter-writing campaign to government she started with schoolchildren brought about the Wild Free-Roaming Horses and Burros Act of 1971. The only way the appalling wholesale roundups and slaughter could be stopped was by shutting down the horsemeat plants. The Wild Free-Roaming Horses and Burros Act is federal legislation and is managed by the Bureau of Land Management (BLM) in the US. As a result of this Act, horses roaming the range increased in numbers and have consequently, and more humanely, been rounded up for adoption.

The adoption of the Wild Free-Roaming Horses and Burro Act in 1971 was a tribute to a remarkable woman, but the method used to stop the inhumane and ill-thought-out roundups was shortsighted. Of course, hindsight is always clear. The resulting Adopt a Wild Horse or Burro program resulted in hundreds of horses once again being rounded up with helicopters and held in corrals for auction and adoption. Well-wishers have bought wild ponies for their kids or themselves, with little or no knowledge about how to finish the horse started by BLM trainers. In any case, from Nevada to Montana and into Wyoming, hundreds of wild horses are held in pens, costing the government megadollars per head. It would seem that the conversation has come full circle: there is talk in some quarters of re-opening horsemeat slaughter plants in several states.

There are, of course, success stories related to the adoption of wild horses in the United States. One of them is Project Noble Mustang, a Montana / Canada mounted border patrol that uses wild horses. Another success story, this one involving wild horses, prison inmates and Homeland Security – an unlikely confluence – evolved under the aegis of the BLM's Wild Horse and Burro Program. In Cañon City, Colorado, prison inmates have been training wild horses since 1986. The training program benefits the inmate trainers as much as the horses. A number of the inmate-trained horses are being used in the Noble Mustang border-patrol program.

❖ ❖ ❖

You have now read about the horses that have gone back to being wild animals of the southeast corner of the Ghost FLUZ. I hope you appreciate the special nature of their interaction with the two wolf packs and other wild animals with which they share their micro-wilderness, which has been protected by its isolation until now. Understanding the value of these rewilded horses is critical to their conservation and that of their habitat. I urge immediate attention by government, industry and the actions of everyday people to bring about the creation of a wild-horse conservation area in the southeast corner of the Ghost Forest Land Use Zone.

Maureen Enns

RIDING THROUGH THE ASPENS.

REFERENCES

Interviews

This book would have been impossible to write without the input of a number of kind people with whom I spoke, corresponded and conducted official interviews. All quotations from interviews are included with permission.

Cori Brewster

Alyssa Butters

Donna Butters

Erin Butters

Erik Butters

Anita Crowshoe

Brenda Gladstone

Lois Green

Rod Green

Hamish Kerfoot

Marina Krainer

Rick Kunelius

Penny MacMillan

Dr. Wayne McCrory

Chuck McKinney

Terri McKinney

Dr. Claudia Notzke

Eleanor O'Hanlon

Dr. Brian Reeves

Darcy Scott

Hank Snow

Jeff Turner

Roger Vernon

Anita Wildman

Cody Wildman

Books

Black Elk. *The Sacred Pipe: Black Elk's Account of the Seven Rites of the Oglala Sioux.* Recorded and edited by Joseph Epes Brown. Civilization of the American Indian series, vol. 36. Norman: University of Oklahoma Press, 1989. First published 1953.

Carbyn, Lu. *The Buffalo Wolf: Predators, Prey and the Politics of Nature.* Washington, DC: Smithsonian Books, 2003.

Chamberlin, J. Edward. *Horse: How the Horse Has Shaped Civilizations.* Toronto: Alfred A. Knopf Canada, 2006.

Cochrane and Area Historical Society. *Big Hill Country: Cochrane and Area.* Cochrane, Alta.: 1977.

Cruise, David, and Alison Griffiths. *Wild Horse Annie and the Last of the Mustangs: The Life of Velma Johnston.* New York: Scribner, 2010.

Dettling, Peter A. *The Will of the Land.* Calgary: Rocky Mountain Books, 2010.

Foran, Max, with Nonie Houlton. *Roland Gissing: The Peoples' Painter.* Calgary: University of Calgary Press, 1988.

Heinrich, Bernd. *Mind of the Raven: Investigations and Adventures with Wolf-Birds.* New York: HarperCollins, 1999.

Lopez, Barry. *Of Wolves and Men.* New York: Scribner, 1978.

McAllister, Ian. *The Last Wild Wolves: Ghosts of the Great Bear Rainforest.* Vancouver: Greystone, 2007.

Mech, L. David. *The Wolf: Ecology and Behavior of an Endangered Species.* New York: Doubleday, 1970.

Mowat, Farley. *Never Cry Wolf.* Toronto: McClelland & Stewart, 1963.

O'Hanlon, Eleanor. *Eyes of the Wild: Journeys of Transformation with the Animal Powers.* London: Earth Books, December 2012.

Purcell, L. Edward. *Wild Horses of America.* New York: Portland House, 1987.

Roberts, Monty. *The Man Who Listens To Horses.* Toronto: Alfred A. Knopf, Canada, 1996.

Sheldrake, Rupert. *Dogs That Know When Their Owners Are Coming Home, and Other Unexplained Powers of Animals.* New York: Three Rivers Press, 1999.

Snow, Chief John. *These Mountains Are Our Sacred Places: The Story of the Stoney People.* Toronto: Samuel Stevens, 1977.

Thompson, David. *David Thompson's Narrative of His Explorations in Western America, 1784–1812.* New edition with added material, edited and with an introduction and notes by Richard Glover. Toronto: The Champlain Society, 1962. First published by the Society in 1916, edited and with an introduction by J.B. Tyrrell.

Uttridge, Sandra, with photographs by Gary Cowan. *The Wild Horses of Namibia*. Cape Town, South Africa: Clifton Publications, 2006.

Watson, Lyall. *Jacobson's Organ and the Remarkable Nature of Smell*. New York: Penguin/Plume, 2001.

Wister, Owen. *The Virginian: A Horseman of the Plains*. New York: Macmillan, 1953. First published 1902.

Articles and Catalogues

Enns, Maureen. *Through the Eyes of the Bear*. Calgary: Art Gallery of Calgary, 2000.

Notzke, Dr. Claudia. *The Status of Wild/Feral Horses in Protected Areas: A Cross-Cultural Perspective, July 1–December 31, 2011*. Unpublished research report. Lethbridge, Alta.: University of Lethbridge, 2012.

Oswald, Paula. "Foreword." In Maureen Enns, *Through the Eyes of the Bear*. Calgary: Art Gallery of Calgary, 2000.

Slater, Dennis. "Diving Deeper: Maureen Enns' Kamchatka Series." In Maureen Enns, *Through the Eyes of the Bear*. Calgary: Art Gallery of Calgary, 2000.

Sniatycka, Ewa. "Curatorial Introduction." In Maureen Enns, *Through the Eyes of the Bear*. Calgary: Art Gallery of Calgary, 2000.

van der Merwe, F.J. "The Real Namib Desert Horses." *SA Horseman* [magazine] (Pretoria, South Africa, ca. 2006).

Wiesenthal, Christine. "Horses of the Ghost." *Lake: A Journal of Arts and Environment*. Issue 4 (Spring 2010). Excerpt accessed October 11, 2012, www.lakejournal.ca/archives_spring2010_christine_wiesenthal.html.

Websites

Alberta:
Galileo Educational Network, www.galileo.org (in the sidebar, under Quick Links, click Wild and Free) [Flash; PDF].

Sable Island:
Sable Island Green Horse Society, www.greenhorsesociety.com.

Sable Island National Park Reserve, www.pc.gc.ca/eng/pn-np/ns/sable/index.aspx.

[All three of these sites were accessed February 13, 2013.]

Rocky Mountain Books
www.rmbooks.com

Library and Archives Canada Cataloguing in Publication

Enns, Maureen
Wild horses, wild wolves : legends at risk at the foot of the Canadian Rockies / Maureen Enns.

Includes bibliographical references.
Also issued in electronic format. ISBN 978-1-927330-24-1 (HTML).—ISBN 978-1-927330-30-2 (PDF)
ISBN 978-1-927330-23-4 (bound)

1. Wild horses—Alberta—Ghost River Wilderness Area—Pictorial works.
2. Animals—Alberta—Ghost River Wilderness Area—Pictorial works.
3. Ecology—Alberta—Ghost River Wilderness Area. I. Title.

SF360.3.C3E55 2013   599.665'50971233   C2013-900443-2

Frontispiece: Pocaterra, silent and immobile, watches us walk by, June 2011.

Printed in China

Rocky Mountain Books acknowledges the financial support for its publishing program from the Government of Canada through the Canada Book Fund (CBF) and the Canada Council for the Arts, and from the province of British Columbia through the British Columbia Arts Council and the Book Publishing Tax Credit.

 Canadian Heritage Patrimoine canadien   Canada Council for the Arts Conseil des Arts du Canada   BRITISH COLUMBIA ARTS COUNCIL
Supported by the Province of British Columbia

This book was produced using FSC®-certified, acid-free paper, processed chlorine free and printed with soya-based inks.

FSC
www.fsc.org

MIX
Paper from
responsible sources
FSC® C016973

Stallions Watching Hope and I, charcoal drawing, 40 × 30 in., Enns, 2008.